KANAKA HAWAI‘I CARTOGRAPHY

Other titles in the
First Peoples: New Directions in Indigenous Studies series

Accomplishing NAGPRA: Perspectives on the Intent, Impact, and Future of the Native American Graves Protection and Repatriation Act
Edited by Sangita Chari and Jaime M. N. Lavallee

Ancestral Places: Understanding Kanaka Geographies
Katrina-Ann R. Kapāʻanaokalāokeola Nākoa Oliveira

Asserting Native Resilience: Pacific Rim Indigenous Nations Face the Climate Crisis
Edited by Zoltán Grossman and Alan Parker

A Deeper Sense of Place: Stories and Journeys of Collaboration in Indigenous Research
Edited by Jay T. Johnson and Soren C. Larsen

The Indian School on Magnolia Avenue: Voices and Images from Sherman Institute
Edited by Clifford E. Trafzer, Matthew Sakiestewa Gilbert, and Lorene Sisquoc

Salmon Is Everything: Community-Based Theatre in the Klamath Watershed
Theresa May with Suzanne Burcell, Kathleen McCovey, and Jean O'Hara

Songs of Power and Prayer in the Columbia Plateau: The Jesuit, The Medicine Man, and the Indian Hymn Singer
Chad S. Hamill

To Win the Indian Heart: Music at Chemawa Indian School
Melissa D. Parkhurst

Kanaka Hawaiʻi Cartography

HULA, NAVIGATION, AND ORATORY

Renee Pualani Louis

with

Aunty Moana Kahele

FIRST PEOPLES
New Directions in Indigenous Studies

Oregon State University Press CORVALLIS

Library of Congress Cataloging-in-Publication Data is available from the Library of Congress.

ISBN 978-0-87071-889-2

∞ This paper meets the requirements of ANSI/NISO Z39.48-1992 (Permanence of Paper).

First published in 2017 by Oregon State University Press
Printed in the United States of America

Oregon State University Press
121 The Valley Library
Corvallis OR 97331-4501
541-737-3166 • fax 541-737-3170
www.osupress.oregonstate.edu

Ka Hoʻolaʻaʻana

(Dedication)

No nā kini akua, for the multitudes of divine entities, who are our original teachers, and all those Native scholars who carry the wisdom of their ancestors with great respect and thoughtfully share that knowledge so another generation may flourish.

Contents

List of Illustrations

Ka Hōʻoiaʻiʻo ʻAna

(Acknowledgments)

This book would never have come into being without the support and encouragement of several people and organizations. It is based on my dissertation, but goes far beyond what I initially set out to accomplish. It could not have been completed without the assistance from The Kohala Center Mellon-Hawaiʻi Fellowship Program and Oregon State University Press First Peoples initiative. Their focus on helping young career scholars publish original research that contributes to and reframes traditional academic silos of knowledge provided the perfect opportunity for the ideas expressed in this book to come to fruition. I would also like to acknowledge the Ford Foundation Dissertation Fellowship and the National Science Foundation Doctoral Dissertation Grant No. 0424892, both of which helped fund my dissertation research, some of which is included in this book.

When it began, this endeavor was based on the learning I experienced in the presence of Aunty Moana Kahele. When she passed, I did not want the connection between those moments of sharing I received in her presence and this moment of sharing I am extending to each reader to be disconnected. So I asked her son, Danford Kahele, for permission to include in this book those handwritten stories she shared with me, in the hopes of reinvigorating the stories and extending their reach.

Young authors such as me are filled with hopeful intent but are rarely capable of seeing the forest for the trees. If it were not for the kind and persuasive nudging of Beth Piatote early in the process, I would not have known how to transform the main focus of my dissertation, Hawaiʻi place names, to the main focus of this book, *Kanaka Hawaiʻi Cartography*. It took an entire summer to strategize the conceptual layout of the book and another three years to complete the research necessary to write it, but I believe it is much better than I had originally conceived. In those three years I came to rely heavily on my many conversations with Kekuhi Kealiʻikanakaʻoleohaililani, without which I could not have expressed the conceptual foundation of Kanaka Hawaiʻi cartography found in this book.

Staying the course is never easy, especially when life conveniently provides ample amounts of distractions. I am grateful for the gentle prodding from my dearest friend, Jay T. Johnson, who selflessly helped provide general course correction to my dead reckoning on several occasions so I could keep my bearings and find my destination.

Writing perfect sentences is beyond my capabilities. Thankfully, I had great help from Everett Wingert, Shirley Bugado, and Noenoe Silva, all of whom took the time to read through various stages of drafts, adding their suggestions and catching some of the glaring grammatical errors. Ev additionally helped tweak both genealogy and both oblique figures. The production process can be quite daunting for a first time author. My experience was nothing short of amazing from working with acquisitions editor Mary Elizabeth Braun and marketing manager Marty Brown to copy editor Susan Campbell and managing editor Micki Reaman.

Lastly, if not for the constant encouragement and support from the three most important people in my life—my parents, Allen and Gloria Louis, and my life partner, Arna Goldstein—I would never have believed I was capable of completing this book.

He Oli Komo

(A Request to Enter)

"Kūnihi ka Mauna"[1]

Kūnihi ka mauna i ka la'i ē
The mountain stands tall in the calm
'O Wai'ale'ale lā i Wailua
Mount Wai'ale'ale in Wailua
Huki a'ela i ka lani
Pulling toward the heavens
Ka papa 'auwai 'o Kawaikini
The Kawaikini ditch
Ālai 'ia a'ela e Nounou
Is obstructed by Nounou
Nalo Kaipuha'a
Kaipuha'a is hidden
Ka laulā ma uka o Kapa'a ē
As well as the expanse above Kapa'a
Mai pa'a i ka leo
Do not suppress your voice
He 'ole ka hea mai ē
There is no answer to my request
(Kanahele, 32)

Welina

(Greetings)

Aloha,

I am Renee Pualani Louis, and this book is a culmination of my evolving understanding of Kanaka Hawai'i cartography. Let me begin with a few clarifications. First of all, I use the term "Hawai'i" throughout this text because the term "Hawaiian" is not a word originating from *'ōlelo Hawai'i*, Hawai'i language. The only exceptions are when I quote authors who use the term "Hawaiian" in their text. Second, let me explain my use of the term "Kanaka Hawai'i" throughout this text.

Hawai'i identity was a dilemma I grappled with for quite some time as I considered using the terms *kanaka*, *'ōiwi*, and *kanaka maoli*. *Kanaka* is explained by Mary Kawena Pukui and Samuel H. Elbert as "human being, man, person, or individual, party, mankind, population" (2003). I decided it was not specific enough, as it could apply to any person or people. Likewise, *'ōiwi*, explained as a "Native" or *kanaka 'ōiwi*, Native person, was also not specific enough, as it could be any Native person (2003). However, *kanaka maoli*, explained as a "full-blooded Hawaiian person," was too specific (2003).

After discussions with Hawai'i political scientist Noenoe Silva, Hawai'i historian Noelani Arista, and *kumu hula*, Hawai'i dance teacher, and Hawai'i scholar Kekuhi Keali'ikanaka'oleohaililani, I decided to use the term "Kanaka Hawai'i," because it combines the genealogical, geographical, and conditional aspects I embrace. Genealogically, it could be a person of Native Hawai'i descent. Geographically, it could be a person from either Hawai'i Island or the Islands of Hawai'i. Conditionally, it could be a person of any descent who embraces Hawai'i cultural practices. This is not an attempt to create another identity category for the people whose genealogy extends beyond the arrival of Europeans to the islands we now call Hawai'i.

I realize this is a very contentious issue. But I can only decide what works for me. I use the term "Kanaka Hawai'i" because it is the best fit for the issues and perspectives I address in this book.

Kanaka Hawai'i cartography is conceptually similar to "performance cartography," first introduced by cartographic historians David Woodward and Malcolm Lewis and defined as "a nonmaterial oral, visual, or kinesthetic social act, such as a gesture, ritual, chant, procession, dance, poem, story, or

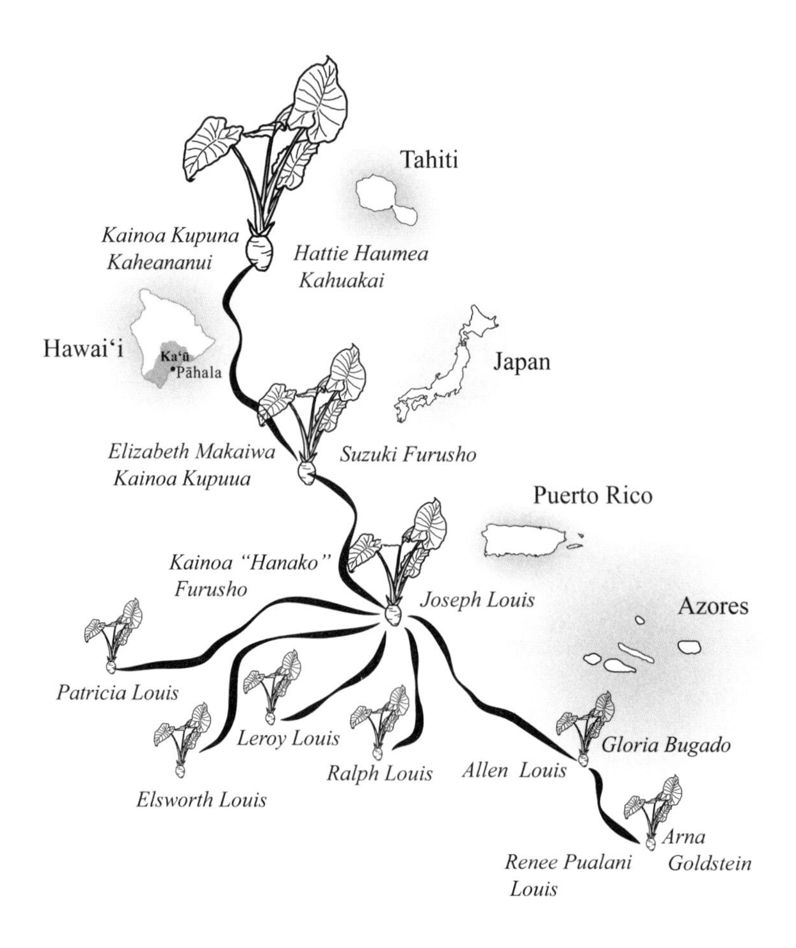

FIGURE 1. Ko'u Mo'okū'auhau Hawai'i (My Hawai'i Genealogy)

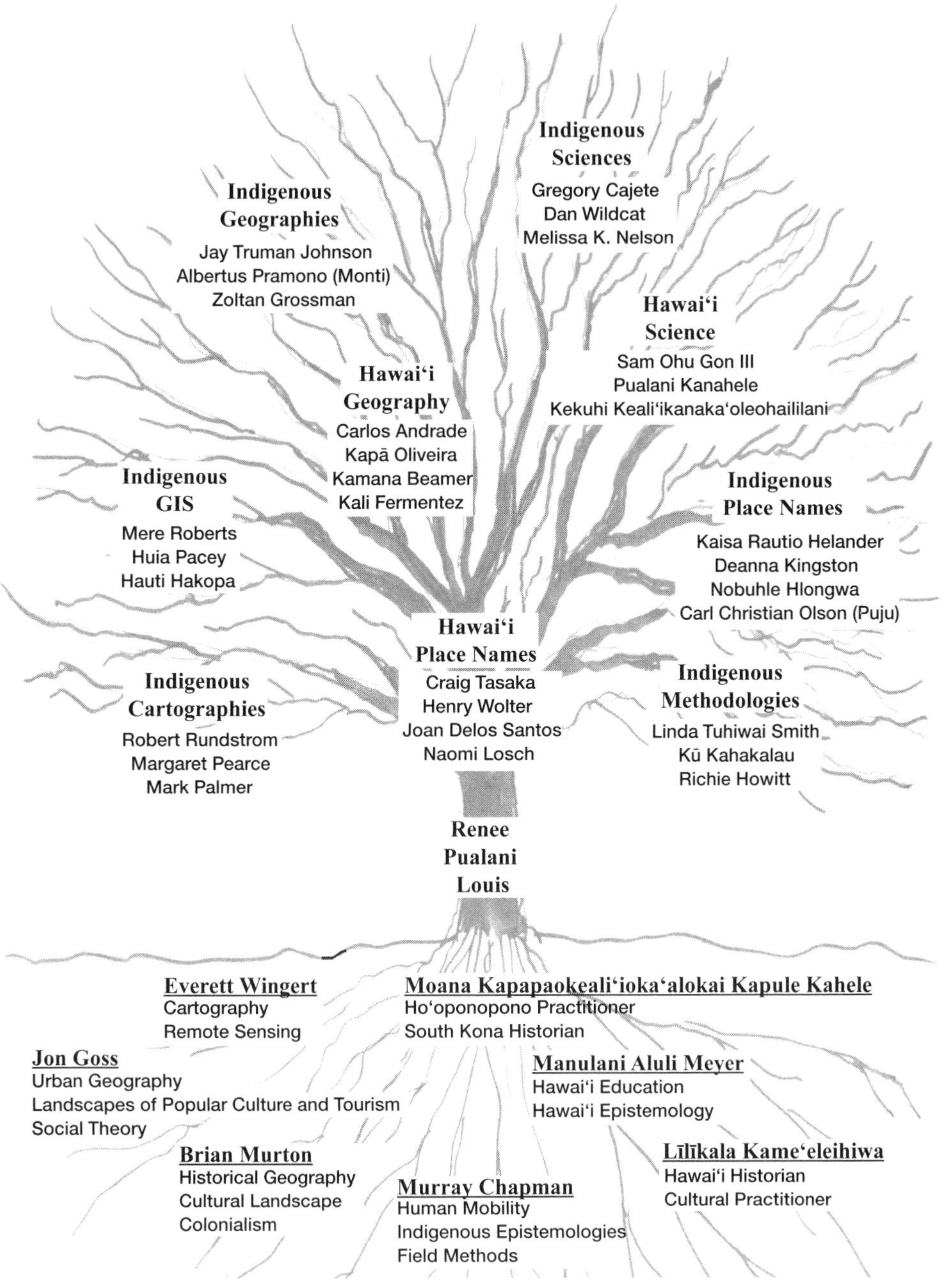

Indigenous Sciences
Gregory Cajete
Dan Wildcat
Melissa K. Nelson

Indigenous Geographies
Jay Truman Johnson
Albertus Pramono (Monti)
Zoltan Grossman

Hawai‘i Science
Sam Ohu Gon III
Pualani Kanahele
Kekuhi Keali‘ikanaka‘oleohaililani

Hawai‘i Geography
Carlos Andrade
Kapā Oliveira
Kamana Beamer
Kali Fermentez

Indigenous GIS
Mere Roberts
Huia Pacey
Hauti Hakopa

Indigenous Place Names
Kaisa Rautio Helander
Deanna Kingston
Nobuhle Hlongwa
Carl Christian Olson (Puju)

Hawai‘i Place Names
Craig Tasaka
Henry Wolter
Joan Delos Santos
Naomi Losch

Indigenous Cartographies
Robert Rundstrom
Margaret Pearce
Mark Palmer

Indigenous Methodologies
Linda Tuhiwai Smith
Kū Kahakalau
Richie Howitt

Renee Pualani Louis

Everett Wingert
Cartography
Remote Sensing

Moana Kapapaokeali‘ioka‘alokai Kapule Kahele
Ho‘oponopono Practitioner
South Kona Historian

Jon Goss
Urban Geography
Landscapes of Popular Culture and Tourism
Social Theory

Manulani Aluli Meyer
Hawai‘i Education
Hawai‘i Epistemology

Brian Murton
Historical Geography
Cultural Landscape
Colonialism

Līlīkala Kame‘eleihiwa
Hawai‘i Historian
Cultural Practitioner

Murray Chapman
Human Mobility
Indigenous Epistemologies
Field Methods

FIGURE 2. Ko‘u Mo‘okū‘auhau Kulanui (My Academic Genealogy)

other means of expression or communication whose primary purpose is to define or explain spatial knowledge or practice" (1998, 4). Although this definition validated the existence of alternative cartographic expressions, it also maintained a Western[2] perspective of spatial reality, which led some authors to disregard Indigenous[3] cartographic expressions as a "display or performance rather than an explanation or record" (Sutton 1998, 365), or as not so much maps as "an oral litany of boundary sites committed to memory" (Mundy 1998, 220). Nonetheless, this validation by Woodward and Lewis was an important shift, allowing many cartographers and social scientists to challenge the philosophical foundations of cartography as a *science of representation.* Most recently, cartographers Rob Kitchin and Martin Dodge proposed another groundbreaking shift, explaining theoretical foundations of cartography as a "science of practices, not representations" (Kitchin and Dodge 2007, 342). This shift provides the platform for the description of alternative cartographic practices that represent alternative spatial realities such as Kanaka Hawaiʻi cartography.

Kanaka Hawaiʻi cartographic practices are a compilation of intimate, interactive, and integrative processes that expresses Kanaka Hawaiʻi spatial realities through specific perspectives, protocols, and performances. It is distinctive from Western cartographic practices in that Kanaka Hawaiʻi recognize the forces of nature and other metaphysical elements as fundamental spatial relationships. Hawaiʻi and Western cartographic practices are not dualistic. Framing them as such is a Western knowledge construct that inevitably places one cartographic practice as the dominant and the other as the marginalized. A Kanaka Hawaiʻi knowledge construct recognizes both cartographic practices as complementary traditions. It is my hope this book provides a conduit for others to define the specificities of their own cartographic practices.

Indigenous scholarship emphasizes the need to position oneself with regard to the research agenda in order to determine the biases and assumptions the researcher brings into the dialogue, description, reflection, and analysis portions of the research. As part of Kanaka Hawaiʻi cultural protocols, sharing a person's genealogy and familial homelands was necessary at the start of certain speaking engagements, allowing the audience an opportunity to find either familial relations or homeland connections to the speaker. It is, in a manner of speaking, a way of establishing both the identity of the speaker and their authority to speak—a way to place the speaker in the scheme of all things. With that said, let me share my Kanaka Hawai'i genealogy and academic training.

I was born and raised on the island of O'ahu and lived the majority of my childhood in 'Aiea. Although I am an only child, I had the good fortune of a close family system and spent many holidays with my cousins, aunts, and uncles. Occasionally I would spend time on Hawai'i Island with my paternal grandparents in Pāhala and my maternal grandparents in Hilo. Members of my father's family are mostly farmers, fishermen, and cattle ranchers, though many no longer continue these professions as their primary source of income. Members of my mother's family were blue-collar workers who augmented their food supply with backyard gardens and took side jobs to supplement their incomes.

I started my master of arts thesis looking into the "lost" ceded lands, hoping to put all the land titles into a geographic information system (GIS) database. After about six months of research, I realized this was not a GIS project. It was a project for an archivist and would take much longer than I alone could carry out. Along the way, though, I read stories associated with various places and ended up writing a master's thesis about the fate of Hawai'i place names on maps. It turned out to be a quantitative look at the decreasing existence of Hawai'i place names on government maps while both non-Hawai'i and hybrid place names increased. Hybrid place names are made up of a Hawai'i place name and a non-Hawai'i descriptor, such as Waikīkī Beach.

My PhD research was inspired by two seminal works, one by David Woodward and Malcolm Lewis on Indigenous cartography, and the other by Hawai'i educator Manulani Aluli Meyer on Hawai'i epistemology. Woodward and Lewis's contribution broadened the scope of what maps represent—from the physical environment to imagined cosmographies—and expanded the definition of cartography, allowing alternative cartographic traditions to be recognized. Meyer's work on Hawai'i epistemology was truly enlightening. It allowed me to frame my understanding of performance cartography from a Kanaka Hawai'i perspective. More importantly, it allowed me to question the reasons why Kanaka Hawai'i, and many Indigenous peoples, prefer performing their understanding of reality. Prior to reading Meyer's work I was convinced Kanaka Hawai'i and Western cartographic practices were dichotomous—natural dualisms of orality and literacy. However, I now realize the very thought of situating these two cartographic practices in a dichotomous relationship maintains a Western philosophical context.

Although these works were inspiring, I had yet to find the question that would direct my research. On the advice of my chair, cartographer Everett Wingert, I opened myself to a moment of inspiration through my dreams. With the guidance of my ancestors, I realized Hawai'i place names were storied symbols in Kanaka Hawai'i cartographies. Furthermore, these symbols, though similar to those found on maps, have more than one explanation depending on the context of the performance. After proposing this theory to my committee in the spring of 2002, I was tasked with finding a "study site." As it just so happened, I had already signed up for a five-week summer course at Aunty Margaret Machado's Hawaiian Massage Academy. *Aia i Hea Au?* (Where Am I?) chronicles my methodological journey, which began in summer 2002 and extended until February 2006.

Finally, please know that the information shared in this text is based on my current understanding of Kanaka Hawai'i cartography according to my own well-informed scholarship. Undoubtedly, by the time this book comes to fruition, that understanding will have evolved. Thus, any errors found herein are mine alone and not a reflection of the sources from which I drew my interpretation. I am sharing my interpretation of Kanaka Hawai'i cartography because I believe cartography has reached a critical juncture in its theoretical evolution where alternative spatial knowledge systems have gained some footing. I am claiming that space for future generations of Kanaka Hawai'i and other Indigenous peoples who have yet to define their own cartographic practices.

RENEE PUALANI LOUIS

Aia i Hea Au? Nānā i ka Wā ma Mua

(Where Am I? Look to the Space/ Time in Front/Before)

One morning in quiet contemplation, I asked the questions, Where am I? Who am I? Why am I here? I received a single response, *Nānā i ka wā ma mua*. The significance of these questions and response lies in understanding the Hawai'i concepts of past and future. The locative term *mua* describes the space/time in front of/previous to the body (Andrews 1865). Conversely, the locative term *hope* describes the space/time at the back of the body. Yet, the Hawai'i language term for the past is *I ka wā ma mua*, the space/time in front of your body, and the term for the future is *I ka wā ma hope*, the space/ time in the back of your body.

There are many ways to interpret this exchange. The only one that matters is what it meant to me at that moment. I had been spinning my wheels, desperately searching for answers to my academic conundrum: good topic but wrong location. I believed the metaphoric meaning of the response I received indicated that I needed to slow down and let the answers reveal themselves to me. I had to get out of my own way and focus on what was really important. My dissertation research was not about me or my academic career. It was about understanding the role place names play in Kanaka Hawai'i cartography. It was with this mind-set that I departed for Aunty Margaret Machado's Hawaiian Massage Academy in the summer of 2002.

Little did I know at the time that I was embarking on the initial stages of what has become known as Indigenous methodology. I, like so many other Native researchers at that time, had read Linda Tuhiwai Smith's book, *Decolonizing Methodologies*, and considered it the authoritative text on conducting research in Native communities. I did not believe I could find all the answers in that book, but it was the proverbial pebble dropped in a motionless pond, generating ripples of ideas, conversations, and controversy. While it did not provide me with a clear path to conducting research in Hawai'i, it did encourage me to remain respectful and accountable, to reciprocate with the people with whom I would work, and to acknowledge the intellectual property rights of the people sharing their knowledge with me.

The remainder of this section chronicles my methodological journey. Much of what I learned is now well documented and accepted as a theoretical research process. Where there is relevance to recent works on methodology by other researchers, I have interjected them into the storied presentation of my journey.

June 2002

Life is different in Kūlou on the Kona side of Hawai'i Island near Kealakekua. I had been living at Aunty Margaret Machado's beach house while taking classes at her Hawaiian Massage Academy. Never heard of Kūlou? Don't worry, you are not alone. Not many people have heard of it. You are not likely to find it on a map as yet, though if I have my way, you may see it on the Hawai'i State Board on Geographic Names website. I first heard about Kūlou the day Hawai'i *ho'oponopono* practitioner Aunty Moana Kahele came to give our class a workshop on *ho'oponopono*, Hawai'i practice of forgiveness and reconciliation.

Aunty Moana was Aunty Margaret's cousin, and I was anticipating a day with less anatomy lessons. I was really hoping she would tell us some stories connected with the place names in the immediate area. After befriending several people living in the area, including Aunty Margaret, her husband, Uncle Dan Machado, and their daughter, our teacher, Nerita Machado, I asked each of them about how and why the places were named, and they all said I should talk to either Aunty Moana or Uncle William Pānui or Uncle Billy Paris.

It has been my experience that most Kanaka Hawai'i know who among them maintain distinct knowledge sets. Some people are quite protective of the knowledge holders in their community and do not respond kindly or directly to questions by researchers. In fact, some US tribal governments and Canadian First Nations have distinct protocols protecting their people and their intellectual property.[4] I recall an archaeologist from the Navajo Nation Cultural Resource Compliance Section saying the Navajo Nation was one of the first tribal governments to create strict protocols for conducting research on the Navajo, especially if a researcher wanted to interview a tribal member.

However, I did not enter this community as a researcher. I did not hide the fact that I was a geography PhD candidate working on Hawai'i place names. I told people who I was both academically and genealogically whenever I engaged in the kind of dialogue people have when getting to know one another. But for all intents and purposes, I was not conducting research. I was a student learning *lomilomi*, Hawai'i massage, who was also curious about the stories of the place names in the surrounding area. I knew that day I would learn about some of them from one of the people many community members respected.

Just thinking about learning the stories made my morning chores go by quickly. Part of the price of staying on-site at the Hawaiian Massage Academy included completing a revolving list of daily chores. It was my turn to clean the yard. I gathered what little leaves were left from the day before then went to get some breakfast. I had the usual coffee, cereal with milk, and toast with peanut butter and jelly. When everyone was done, we all pitched in to set up our classroom on the front porch, rearranging our breakfast tables and chairs into the typical classroom setup.

Nerita was always on time, but Aunty Margaret and Uncle Dan came and went, depending on how they were feeling that day. We later learned that Aunty Margaret was much more interested in the *lomilomi* lessons that started in the afternoons of week two than in the anatomy lessons that we all had to take as part of our licensed massage therapist (LMT) state certification. Our anatomy teacher was a former student from Maui, Dr. Mark Lamore. He made sure we were well-versed in all the areas the exam would cover regardless of whether we decided to become LMTs. Every now and again Nerita would take time to read a story from one of her Seventh-Day Adventist books to provide the spiritual aspect of our learning. This was one of those days.

The lesson Nerita was sharing that morning was overshadowed by the mosquitos making a meal of me. The summer heat brought lots of them, and my sweet blood made a scrumptious target. I remember thinking, as Aunty Moana arrived, I had to get the so-called good stuff, with that nasty chemical, because the repellant I bought from the natural food store was not cutting it. The driver parked on the grassy knoll next to the house and pulled Aunty Moana's wheelchair out of the trunk. We all went down to greet her, and the men in the class thought out loud about the best way to get her up the eleven steps to the patio.

Once we were situated and all of the niceties were out of the way, we started our lesson on *hoʻoponopono*. Although information about *hoʻoponopono* is readily available online or in many books today, we did not have that luxury in 2002. I had heard about the practice but had never personally participated in the process. I knew there was no such thing as "one" correct way to practice *hoʻoponopono*, as each practitioner did things according to the way they were taught. Nonetheless, I was eager to learn from Aunty Moana.

She had a few worksheets for us to look over but knew her material so well she rarely referred to them. She began humbly, telling us about her life, her family history, and genealogy, as it is a common practice to establish connectivity and authority on a subject before speaking about it. She was born in Nāpoʻopoʻo, Kona Hema, Hawaiʻi, and raised by her paternal grandmother, Lokalia. She told us how she was chosen to learn *hoʻoponopono* at the end of a ritual her grandfather, John Au, made her and her cousin(s) complete. She wanted to learn *lomilomi*, but Aunty Margaret was chosen to learn that and Aunty Moana was taught *hoʻoponopono* and *lāʻau lapaʻau*, a Hawaiʻi practice of addressing pains and illnesses with plants and prayer. Aunty Moana did get to learn some *lomilomi*, and Aunty Margaret learned some *hoʻoponopono* and *lāʻau lapaʻau*, because each cultural practice relates to the other.

Aunty Moana said, for her, *hoʻoponopono* is a method of releasing the negative that resides within, whether we know it is there or not. She made it very clear that *aloha*, meaning compassion, sympathy, kindness, grace, is the key to this method, and that it is very big work that must be approached with respect. It could not be rushed. She said people protect their pains, and as we work on their bodies through *lomilomi* we could also be stirring up emotional issues. If we recognized this was happening, it would be best to recommend they get some help from a *hoʻoponopono* practitioner. She explained the basic steps of *hoʻoponopono* and gave us examples from her experience as a practitioner. Her intent was to make us aware of the effect our own energy could have in working with clients and vice versa. She wanted us to learn how to release any unhelpful energy that built up in any of us in order to protect our clients and ourselves.

The one thing she said to do every day without fail was to sit as the sun sets at the end of the day and release all the unhelpful energy. We were told the hardest things to let go of were fear, anger, shame, and guilt. It might take many days, weeks, or even months for us to get through those emotions, but we would better understand ourselves, and as a result we could be of better service to our clients.

When Aunty Moana was done speaking, she answered our questions, and when it appeared as though the questions about her teaching were nearing an end, Nerita asked if she would tell us some stories of the place names around this area, because there was a student interested in hearing about them. It took her a moment or two to gather herself. She began with the story of Kūlou, which was on the shoreline almost directly in front of our classroom. We also learned about Kapahukapu and Hali'ilua and Kealakekua. But perhaps the most important thing I learned was that the bay was not named Kealakekua before Captain Cook came and put it on a map. According to the people she heard growing up as a child, the name of the bay was Kapukapu.

She spoke until it was time for us to break for lunch. Our morning session went longer than usual, but it is considered rude behavior to stop or interrupt a *kupuna*, Hawai'i elder, when they are talking. So, we got an extra half hour for our lunch break. Most everyone took that as an opportunity to go swimming first and then grab a quick lunch. I, on the other hand, preferred to stay nearby, in the off chance I could talk with Aunty Moana during lunch. I was hoping to ask her if she would consider working with me.

She spent most of the lunch break talking with Aunty Margaret, Uncle Dan, and Nerita, catching up on family stuff. I was far enough away not to be eavesdropping but close enough to try to catch her attention before she left. In other words, I made it obvious I wanted to talk with her, but I did not want to intrude on their space. Nerita called me over and told Aunty Moana that I was the student asking everyone about the stories. It is far more beneficial to have someone introduce you to a *kupuna* than it is to just walk up and state your intent and interests. That kind of "cold call" approach usually sets the tone of the conversation and likely garners an equally chilly response. However, an introduction by a family member indicates a certain level of trust has already been achieved, depending on the family member, of course. Since Nerita was a practical and sensible woman, Aunty Moana received me warmly and asked me why I was so interested in that stuff.

I told her I believed Hawai'i place names held more information about the place than just a label on a map could share. She agreed and spoke about how so many names on maps were wrong and that if nobody does anything about it, the next generations would be cut off from that information. I asked if I could call her and talk with her about my project and see if she wanted to help. I was pleased that she gave me her number and told me the best time to call was in the evenings.

I do not remember what happened for the rest of the day. I felt like a path had been presented before me, and I was both anxious to get started and frightened about what I had gotten myself into. I received no specific training on how to proceed. Linda Tuhiwai Smith writes, "Most indigenous researchers who work with indigenous communities or on indigenous issues are self-taught, having received little curriculum support for areas related to indigenous concerns" (1999, 135). At the time, the University of Hawai'i did not have curriculum designed specifically to help me understand where to go from here.

Thankfully, I had committee members who made access to that information easier for me. However, it was summer, and they were on another island. So, I did the only thing I could think of. I sat on the rock near the shore as the sun went down and asked *nā akua*, the multitude of divine entities or processes, to help me shed my fears, my pride, and anything else that would limit me from doing what was appropriate and necessary to learn the stories in a manner that honored the intelligence of the generations of people that had lived in this place.

April 2003

A blast of warm air rushed across my body as I opened the door from the Hawai'i State Archives building. I immediately stepped away from the door, put my bags down, and removed my jacket. Why does the building have to be so cold? Okay, okay, I know it is to protect the documents from bugs and deterioration, but geesh, the temperature difference was enormous. I was perspiring before I got back to my car. Ten months had passed since I had met Aunty Moana, and I had spent the better part of it looking up, scanning, photographing, and recording place names for the area surrounding Kapukapu. I found a total of 296 place names from both map documents and descriptive sources.

I decided to use the area Aunty Moana was most familiar with as my research area. I had maintained contact with her on a regular basis via phone conversations. I did not learn about maintaining regular contact with community participants from any class I took at the university. It was something I learned at home, a common courtesy taught to me by my parents. I always started my call with, "Aloha Aunty Moana, this is Renee, the girl doing the place names research." After the first two phone calls, she really did not need the introduction, but I continued to do it as sort of a ritual.

I spent a couple of months in each of the main offices housing historical documents and called to let Aunty Moana know about my progress on finding archival documents with place names on them. Sometimes she would tell me about other places to seek information, but most of the time she would tell me stories of those place names she had told the massage cohort during the previous summer. The stories did not vary much at all. It got to a point where I could recount the stories very well and knew what was coming next.

I planned a few one- to two-day visits with her as often as I could, sharing lists of archival documents I found at the Hawaiʻi State Archives or Hawaiʻi State Survey Office in Honolulu so we could prioritize the order in which I should acquire the documents. But more importantly, I brought her fish and poi or different native plants from my backyard that she did not have in her potted plant collection. This sharing was not about my generous nature. It was something different. I did not realize it until sometime later when I was reflecting on my time with Aunty Moana, but it was almost like a rite of passage. Aunty Moana would say how much she liked some food from a certain store or person or give me pointers about specific types of native plants, and I would notice that she did not have them in her collection. I made a mental note of those things and attempted to acquire them. I was not always successful, but interacting with people in an attempt to find those items added more flavor to the sometimes bland process of data gathering.

This kind of reciprocity and data sharing has been identified by numerous researchers in a wide range of disciplines[5] and is included in the Association of American Geographers Indigenous Peoples Specialty Group's "Declaration of Key Questions about Research Ethics with Indigenous Communities." However, as I did not have these resources at my disposal, the key for me was to be genuine. Bringing food or plants to Aunty Moana was a natural extension of who I am as a person. I was not going out of my way to do what was necessary to please Aunty Moana so she would like me better or trust me more. I was doing it because I was raised to bring something for the person with whom I planned to visit, especially if I was stopping by their home. Sharing the research process was a natural extension of that reciprocity. It was my opportunity to show how much I was invested in learning the place names in the area surrounding Kapukapu and how much I respected Aunty Moana's opinion about the archival work I planned to do.

Since she had not seen any of the documents I had on the lists, she asked if I would bring them all. I acquired, scanned or photographed, and transcribed over a hundred pages of handwritten documents, color-coding all the Hawai'i words in descriptive texts such as Boundary Commission Testimonies in order to distinguish between people's names, place names, and general geographic terms like *mauka*, toward the mountain. The Boundary Commission Testimonies are handwritten records of land boundaries transcribed from oral accounts of community members. For those of you unfamiliar with Hawai'i history, in December 1845, Kamehameha III passed Hawai'i's first Organic Act. Two months later, in February 1846, the executive branch passed the second Organic Act, which provided regulations for its departments and, most importantly, established the Board of Commissioners to Quiet Land Titles (the Land Commission).

The Land Commission was to determine existing land rights as of February 1846. The claim process involved four steps: (1) present a claim, (2) procure a survey, (3) appear before the Land Commission to testify on behalf of the claim, and (4) submit approval from the Land Commission to the minister of the interior to receive a Land Court Award (LCA), a Royal Patent (RP) or a Māhele Award upon payment of a commutation to the government. Payment was typically one-third the value of the land you claimed. If you did not have the money, you could either pay with a portion of your land or lose your right to claim any of it.

I knew there would be redundancies—same names with different spellings—between the different sources. So, I used a spreadsheet to keep track of everything. I sorted the list according to place name and then according to source and gave preference to descriptive sources because they were chronologically older and closer to the primary source, the people living in the community. Of the 296 place names, I only considered 162 of them unique—132 from descriptive sources and thirty from map documents. I was amazed at the number of place names that were not on maps. I learned during my master's research that there would be more place names in the descriptive sources and knew that cartographers make decisions on which place names to include on maps. Furthermore, since the maps I found were all made by or for the government, I knew the place names would be associated with land divisions. I had no idea there would be over four times as many unique place names in descriptive sources.

After several phone conversations and short one- and two-day visits, I was excited to spend the entire summer of 2003 in Kona visiting with Aunty Moana and other community members. I brought everything, from USGS topographic maps to copies of documents, pictures, and maps I had found in the libraries, Hawai'i State Archives, Hawai'i State Survey Office, and the Bishop Museum Library to my transcribed work, to facilitate our open-ended discussions.

The Boundary Commission had twenty-one testimonies for my area of interest, with eighty unique place names, seventy-nine of which had no known coordinates. I hoped Aunty Moana could help me locate some of these places, as all my efforts to locate them were unsuccessful. I even acquired a digital video camera and a digital audio recorder with lots of tapes to keep a record of my conversations and reflections. When I gave Aunty Moana copies of all the work I had amassed, she carefully reviewed everything and said, "This is important work."

Summer 2003—Research Consent Forms

Of course I knew I could not "officially" engage in research without receiving consent from Aunty Moana. This sticky point really concerned me, because it felt wrong. I did not know why at the time but knew enough to be apprehensive about bringing out a piece of paper that Aunty Moana had to sign in order to make it okay for me to "collect research." I had no idea how to approach Aunty Moana with it. It seemed foreign compared to the kind and style of conversation we were having, and I did not want to change that easy nature. So, I included it as one of the documents I shared with her on one of my short visits. I explained that the University of Hawai'i gave me a human subject research exemption because I was not going to work with children, nor was my project medical in nature. But I still had to write an Interview Consent Form and get everyone I spoke with to sign it. I asked Aunty Moana if she would help me with the draft of the Interview Consent

Form I was working on and handed her a copy. She took the draft, promised to read it, and continued with the earlier conversation.

I visited a couple more times and did not mention "the paper." I knew from Linda Tuhiwai Smith that gaining informed consent could take several months if not years and that "consent is not so much given for a project or specific set of questions, but for a person, for their credibility. Consent indicates trust and the assumption is that the trust will not only be reciprocated but constantly negotiated" (1999, 136). I trusted that Aunty Moana would bring it up when she was good and ready. However, the time was nearing when I was going to have to broach the subject or knowingly walk into a gray area of research ethics. So, I waited for the right moment and chose to tactfully bring it up when she was talking about someone "taking what she said and publishing it without her knowledge." That was my moment.

I told her the Interview Consent Form I gave her a while back guaranteed that would not happen with the work we were doing. She told me she had read it and was all right with most of it. She had signed many consent forms with other researchers, and it was not a problem. I was curious about the parts she was not comfortable with and asked her what I could do to make it better. She said the signed form means more to the agency than it does to her, and she believed after working with me for several months that I would "do the right thing" by her and her family. The moment she uttered those words, I realized the difference between the ethical documents the university deemed necessary to conduct research and the moral responsibility I had, not just to Aunty Moana, but to her family. This was an important turning point in my research. This is the moment I began advocating Indigenous research methodologies.

I finally understood what Hawai'i historian Lilikalā Kame'eleihiwa meant when she said, "Being born Hawaiian means you are born with responsibility." At the time I heard the statement I believed it had something to do with activism. Indeed, for some people it does have to do with activism, but for others it means maintaining cultural protocols and practices. For me, in this instance, it meant merging personal morals with codes of ethics when doing research. Any research involving Indigenous peoples always has a political element. Even my seemingly benign interest in Hawai'i storied place names had a downside. For example, developers interested in increasing tourism could take the research and misrepresent the stories in order to provide a particular area with a more "authentic" Hawai'i connection. There is also the possibility of a story revealing the location of scarce resources or providing a clue to those seeking artifacts. These scenarios could ultimately result in corporate lobbyists pressuring government entities to change zoning boundaries or enact laws in favor of their economic or personal gain. As a Native researcher, I needed to make sure Aunty Moana knew the possible problem areas publishing the stories could provide.

As we spoke about the specifics of these issues for a few more minutes, she also revealed that many people came to speak with her about the old ways but only a few acknowledged her, fewer still offered to pay her, and almost all of them forgot about her when they left. She said very few people told her what they were going to do with the information she shared, and so she rarely knew the extent or distance her words traveled. That bothered her quite a bit. She said she shared specific information with people based on what they were going to do with the information. It was sort of like different levels of security clearance. There was some information that everybody could know and some information that only a select few were ready to hear.

I understood immediately that the context of when and where information is shared mattered greatly. It really should not have surprised me. After all, doctors do not reveal a patient's status to just anyone, nor do other respected professionals in their specific capacities. Yet, for some reason, prior to this discussion I had not given a knowledgeable and respected community member the same consideration. There was no way I could I have known this. I was taught that researchers are rewarded for their objectivity and that it was not necessary to form any kind of relationships with research participants. In fact, academia deems the research an academic conducts to be the intellectual property of the researcher, thereby allowing researchers to use the results as they deem fit, including presenting them to an audience with whom they were never meant to be shared or at a time that is not appropriate.

Aunty Moana certainly had a good point: not all information is meant for everyone. The only way I would know if it was appropriate to share the stories she shared with me was to ask her every time I wanted to share them. She needed to know who the audience was, where the presentation would be given, and the extent to which the information could possibly travel. She also needed to know that I would agree to her decision on whether or not she thought it was okay to share. I consented to consult her prior to any presentation I made, oral or written, regarding the work we were doing together, and to acknowledge her as a collaborating participant.

Summer 2003—Dreams

I conducted our discussions in English, and at times, Aunty Moana would say a few phrases in Hawaiʻi language, asking if I understood what she said. Most of the time I did understand her, but responded with my limited language skills and usually finished off in English. This naturally set the tone for our rapport. It did not stop her from using Hawaiʻi language phrases; she just did not expect me to respond in Hawaiʻi language. I am certain that a researcher with more confidence in their Hawaiʻi language skills would have received a different depth of sharing to which I was not privileged.

Although I initially set out to digitally record our discussions, both audio and video, Aunty Moana preferred that our sessions be conducted without being taped. Instead, she referred me to another video done by Kamehameha Schools Land Asset Division. In one instance, I was told specifically that the stories being shared were for me to remember and not for others to know on a tape. I was able to get a copy of both the report done by Kumu Pono Associates and the video done by Nā Maka O Ka ʻAina.

As Aunty Moana spoke of different places, she weaved in genealogies and personal experiences. After about three or four sessions, I started asking if the person she was talking about was the same one from a different story, stating, "You know, the one where...", finishing the sentence with the story she had told me. Occasionally, she corrected me, emphasizing points I was certain I hadn't heard before. It was about this time that I shared with her two of the dreams I had about Kealakekua and Kapukapu.

Although having a well-known and very deeply respected person willing to share the stories of Kapukapu could be seen as reason enough to conduct this research, I acknowledge that, for Native people, the search for knowledge is much more than a physical task. It is also a spiritual learning. I continue to hear many stories of academics doing research on Indigenous people or in Indigenous communities for research's sake, without bothering to ask if either the people or the place could benefit from the research.

I didn't want to be one of those people, and while I knew Aunty Moana wanted to work with me and wanted to share her knowledge, I just wasn't sure the place was ready to share itself with me. So, I did what came naturally. I prayed for a sign, a vision, a dream, anything that would let me know that it was *pono*, meaning proper, righteous, virtuous, to do this research and learn about the intimate details of Kapukapu. Thankfully, the answer came in the form of several dreams over several weeks. These are two of the dreams I shared with Aunty Moana.

In the first I was a young girl running, playing with a young Kanaka Hawai'i boy in a forest. I was chasing him on a worn path, both of us laughing as we took turns hiding with the chance of surprising the other unawares. We decorated each other with flowers and ferns picked along the way. We climbed trees, made birdcalls, picked and ate fruit. By and by we reached a community near the shoreline. I observed many structures, men fixing fishing nets, women picking seaweed. Although the people noticed us, they returned to their work speaking to each other . . . in Hawai'i language.

That was certainly strange, because although I passed my language requirement by taking a third year of Hawai'i language, I still didn't feel confident in my ability to speak Hawai'i language and was still only somewhat confident in my ability to comprehend oral communication. So, hearing them speaking Hawai'i language and understanding them in my dream was strange, in retrospect. Nonetheless, no one stopped either the boy or me from playing. No one scolded us for making too much noise. No one warned us not to go over the ridge.

I remember feeling like this was a new place to me. Like I was the new kid and this young boy was from this village. I felt that he must know the right places to go and not to go, because I certainly didn't feel like I knew. We continued to play, winding our way up the mountain playing hide-and-seek in empty caves. The higher we got, the funnier the air began to taste, and all of a sudden we were at the ridge, and we were very quiet. The little boy's eyes were in distress as he motioned for me to meet him on the summit. I slowly joined him and saw the reason for his anguish. On the other side of the ridge was the modern development Kona has become, with houses, roads, industrial warehouses, and an ocean filled with motorized fishing boats. Gone were the trees, gone were the birds, and gone were the places for the practices of the people of old.

Without saying a word, he communicated to me with a long hard glare, and I knew what I was supposed to do. As he returned to his village, his time, I stood and walked the summit toward the ocean, the village on one side, the modern development on the other. These two incongruent cultural landscapes were separated by this ridge that I walked like a fence until I reached the ocean. I sat on the cliff pondering what this all meant as the sun began to set. I realized I was at a juncture in space/time. I knew what was coming for the village. I knew the cultural landscape that celebrated a Hawai'i understanding of life would soon be engulfed and could quite possibly be forgotten . . . unless some people chose to remember, and not just remember but remind others of what existed here before it became thoroughly swallowed up.

Becoming or being one of those people is a tall order, a hefty responsibility that I was not sure I was chosen to carry out. So, I did the unthinkable. I dove off the cliff into the ocean below, even though I somehow maintained my fear of the ocean in the dream. I remember thinking that if it was the right thing to do, I wouldn't die. I had heard somewhere that if you die in your dreams, you are quite possibly dying, and I didn't want to die. I just wanted to know for sure that this was right.

I didn't die. I remember surfacing rather relieved. I was treading water when I saw probably my greatest fear approaching, the huge dorsal fin of a shark. I began thinking now I was going to die. But then I didn't panic or feel like fleeing. I remember thinking if this was it, there was nothing I could do. Then I realized I was surrounded by all the creatures in the ocean, turtles to the left of me, dolphins to the right, and various fish scattered between them, including rays and eels. As the shark slowed and settled in front of me, the circle was complete. The awe I felt for all these ocean creatures to surround me as such was so great, the meaning too much for me to comprehend, the power too immense to perceive. I awoke, but as I did I remember looking down seeing it all: the circle, the village, the modern development, and me.

By the time I had my second dream, I had a chance to speak with someone about my fears of the ocean. I expressed the reasons and the rationale for my fear and was given very good advice from committee member Manulani Aluli Meyer in a personal email communication. She said fear lives in our minds. When it is shaped by experience, it becomes a conception that is difficult to change by just thinking about it. To remove this kind of fear, I had to get out of my mind and return to my body. I had to go into the ocean and deconstruct the fear I had created in my body by retraining my body to change my response to the ocean.

I did just that. I started taking small steps to get over my fear and began visiting the ocean more often and staying in the water longer with each visit. At first the slightest touch of any object would send a bolt of terror through my body. Eventually, I got over the terror of things touching me in the ocean and started working on feeling comfortable treading water. This is a good place to tell you about the second dream, because it starts with me treading water in the middle of Kapukapu.

In this dream I have no idea how I got to be in the middle of the bay. There are no boats around me, no kayaks, and no people, just the calmly lapping sounds of the ocean all around me as I tread water looking west into the Pacific Ocean. In this dream I am not afraid of the ocean or of not being able to feel the earth under my feet. As I turn to my right, I see the flat of Ka'awaloa and imagine the *ali'i*, Hawai'i leaders, that made their residences there. It seems like an excellent place for affairs of the government. It's near a permanent aquacultural food supply, has access to agricultural fields up *mauka*, has easy, quick access to launch an attack or flee from one, and has several brackish water holes.

As I continue to turn to my right, Kapaliomanuahi rises from the Ka'awaloa flats and becomes Nāpalikapuokeoua. The sight is immense, and I realize that where I'm treading water, the ocean floor is probably as deep as those cliffs are high. I continue to turn to my right, facing east looking toward the beach. I imagine the shore once lined with sand and small structures for the *kāhuna*, master practitioners, who lived and practiced here. Still turning to my right I see Hikiau Heiau and realize it would have been the tallest structure on the beach, but is now dwarfed by modern homes that continue to line the coast as I turn to the south. It is at that moment that I sense the presence of another. It was the shark, Kua, from Ka'ū, an ancestor for many families from Ka'ū and the namesake for Kealakekua, according to Aunty Moana's story.

I turn to face him. It seems as though I know that this is the reason I was there treading water. I was waiting to meet him. In retrospect I am really surprised I was not afraid of treading water or the arrival of a shark or meeting such an important ancestral entity. I turned completely toward Kua and said, "Ah, there you are." He swam by, nudging me, and I took it as a sign to hold on, which I did. It's amazing that you can breathe underwater in your dreams. He gave me a tour of the bay, showing me the many crevices and underwater caves. It was beautiful. When it was time to go, he looked me in the eye. I recognized that look. It was the same glaring look the little boy gave me at the summit. They were one and the same.

It was after this dream that I finally felt this research was the right thing to do. I, of course, shared these experiences with Aunty Moana, and our talks became more intense. She would still take quite a few minutes talking about the demands others were placing on her time, but she more quickly moved on to telling me personal stories and experiences she had in connection to the spiritual landscape of Kapukapu and its surrounding areas. I literally felt myself transported into the stories she told—and I didn't even have to close my eyes to imagine them. As I sat on the floor of Aunty Moana's living room, I had no idea there was actually a pattern being revealed. She began with the story of Kealakekua as she had heard it passed down from generation to generation. She then elaborated on the misrepresentations of seven names that have been changed and circulated in textual and cartographic sources. Lastly, she breathed life into Kealakekua, revealing nine intimate stories of sensual geographies.

She would always give me a few days between sessions and encouraged me to spend that time in those places I learned about. She said we remember better when all our senses are engaged in the learning process. That way the place and the story fused with our *naʻau*, small intestines, and metaphorically, the seat of thought, intellect, affections, and moral nature (Andrews 1865). Our minds record everything, but our recall is usually limited to those things on which we focus. By experiencing the world more sensually, we allow our minds to make more subtle connections with each place. A practiced mind associates smell, temperature, humidity, wind direction, and the rhythmic movement of plants with the precursor of bad weather and will automatically bring a jacket to work without even thinking twice.

I spent many, many hours learning dozens of stories from Aunty Moana that summer. Stories that became a part of my being as I sensually experienced as many places as I could access. I was sad for my summer of learning to come to an end, because I knew I had only scratched the surface of knowledge maintained by this respected community elder. After each one of our discussions I wrote down as much as I could remember in personal journals. However, I remember thinking that I didn't want my writing things down to take away from the experience of remembering the narrative, from experiencing the performance. I reminded myself that these notations were not meant to be a substitute for the performances. They were meant to aid me in writing this manuscript.

On my last day that summer with Aunty Moana, I wrote down the place names of all the stories she had shared with me and asked her which ones I could share in my dissertation. We discussed the reasons for each selection, and just before I left she handed me a handwritten copy of her manuscript, "Clouds of Memories." She explained that she wrote all these stories down as she grew up listening to friends and family "talk story." Since it was her last and only copy, I refused to take it and later acquired a copy from a friend to whom she had given a copy some years earlier.

I continued to call and visit Aunty Moana regularly after I returned to O'ahu. The phone calls became fewer and farther apart, mostly because she was very busy working on other community matters. However, I was beginning to write up the work we had done together and wanted to present it at different conferences and needed to share that with her before turning in an abstract. When I did get through to her, she asked me where the conference was being held, who would be attending, and why I thought it was important to share the information. I asked before each conference and each article that referenced any information she shared with me. Although these conversations were the most tedious, they were also the most liberating because I knew I had received her blessing. This may seem excessive from the perspective of an academic code of ethical conduct, but it was the right thing to do. I continue to honor Aunty Moana at any presentation that contains any information from our work, or rather my training, together.

Sometime later Aunty Moana was checked in to the Kona Hospital when she could no longer care for herself. Thereafter, I visited with her at the hospital, bringing her my latest chapters or articles. By then, she had grown to trust my representation of the information she shared with me, and most of our time together was filled with retelling stories. In the last few months of her life, I discovered someone had stolen Aunty Moana's handwritten copy of her then-unpublished manuscript.[6] On what was to be my last visit with her, she asked me to make copies of the manuscript I had acquired from a friend so she could give them to her adopted son and daughter. I gave her my "field" copy that day, knowing I had another "clean" copy at home. I told her I would bring the other copy next time, when I got back from a conference. At that moment, I realized there would not be a next time.

That was one of the hardest visits of my life, second only to being present at the death of my maternal grandmother. I asked if I could scan and include some of her stories in my dissertation, and she agreed to six of them. Of the twelve other stories she shared, she agreed I could transcribe seven of them from her then-unpublished manuscript, "Clouds of Memories." She asked me to recount the other five stories from my recollection and, after careful consideration, agreed I was ready to share them from my own voice. Aunty Moana may not be in this world, but she is still part of my reality. I know those stories I share from my own voice carry with them the weight of her ancestors. To this day, I do not share them without first asking permission. I did not want to leave her bedside. I did not want this journey to end. But the nurse came in and said visiting hours were over. I kissed Aunty Moana goodbye on the cheek, and told her I would remain "open" to her continued guidance. She was calm and composed and graciously reminded me, "This is important work."

Ka ʻŌlelo Mua

(Introduction)

> Land, body, and memory all inform one another. The land, extending out and into the ocean, holds the practical and epistemological memories of encounters. The body is the agent, the participant in the environment, and the container for memories. For Hawaiians in the past, vital information was relayed through the environment, and this memory of *ka ʻāina* (the land, that which feeds) affected close interpersonal relationships and societal structures.
>
> —S. L. Iaukea, *The Queen and I*

Experiencing a place without sensual inhibitions reawakens a primal connection with the elements of nature. It provides opportunities to resynchronize our internal clock with the rhythmic cycles of the seasons and to reconnect us with our responsibilities as caretakers of this planet. Although Hawaiʻi has been an occupied colony for over a century, many Kanaka Hawaiʻi still chant to the rising sun, pray before entering a forest, plant and harvest by the moon, fish for specific species based on certain flowering plants, and honor ancestors who halted warfare based on the location of Makaliʻi, the star constellation also known as the Seven Sisters or the Pleiades. This sensual engagement with nature is the *niho pōhaku*, tooth stone used to build interlocking rock foundations, of Kanaka Hawaiʻi cartography.

Kanaka Hawaiʻi cartographic practices are a compilation of intimate, interactive, and integrative processes that express Kanaka Hawaiʻi spatial realities through distinct perspectives, protocols, and performances. Many Kanaka Hawaiʻi still perceive themselves as an extension of nature and treat all natural and metaphysical elements as part of a sacred genealogical relationship. Precontact Kanaka Maoli incorporated vital spatial knowledge about the places where they live, work, and pray into various cultural practices, including *ka hoʻokele*, Hawaiʻi navigation; *ka haku ʻana*, Hawaiʻi verbal arts composition; and *ka hula*, Hawaiʻi dance.

The cartographic natures of these well-known Hawaiʻi cultural practices have been long overlooked. When Kanaka Hawaiʻi embraced Western cartographic techniques and technologies to represent Kanaka Hawaiʻi spatial knowledge in order to adapt to new power regimes, many people believed those modern representations were the extent of Kanaka Hawaiʻi spatial knowledge. But Kanaka Hawaiʻi never discarded their own cartographic practices. Kanaka Hawaiʻi cartographic practices have been sitting right before our eyes, waiting for the right time to be recognized and appreciated.

When I say it has "been sitting right before our eyes," I am referring to *ka wā ma mua*, the space/time in front/before. Lilikalā Kameʻeleihiwa was the first Hawaiʻi historian to write on this concept embedded in the Hawaiʻi language. She states that, "It is as if the Hawaiian stands firmly in the present, with his back to the future, and his eyes fixed upon the past, seeking historical answers for present-day dilemmas" (1992, 22–23). This very important philosophical concept is indicative of how Kanaka Hawaiʻi value the knowledge cultivated by Hawaiʻi ancestors. Many Kanaka Hawaiʻi still look to the past when searching for answers, thereby maintaining intimate connections with their ancestors through ritualized prayers and other practices. This does not mean that Kanaka Hawaiʻi should not or do not engage in technological innovation.

Many Kanaka Hawaiʻi have adapted to new technologies without losing perspective of their responsibility as caretakers of their *kulāiwi*, homelands. For example, many Kanaka Hawaiʻi free-divers use modern spearfishing technologies, elastic-powered spearguns and slings, or compressed gas pneumatic-powered spearguns, but they observe traditional *kapu*, prohibition, on those fish varieties that are mating or spawning, and they are vigilant not to overfish or strike at baby fish. Obviously this is not true for all Kanaka Hawaiʻi, as it is nearly impossible to speak in absolutes about any group of people.

On that note I would like to clarify that in sharing my understanding of Kanaka Hawaiʻi cartography, I am identifying those Kanaka Hawaiʻi spatial knowledge systems embedded in a broad range of historically recorded and accepted Hawaiʻi cultural practices and describing how these practices solve relational spatial/temporal problems from a Kanaka Hawaiʻi perspective. However, these cultural practices are not absolutes either. Each island cultivated specific techniques that reflected the resources, technological ingenuity, and natural features of their *kulāiwi*. Therefore, wherever possible, I identify the origins of a specific cultural practice to the best of my knowledge and indicate whether or not there are differences elsewhere in the Hawai'i Islands to the best of my knowledge.

When I say Kanaka Hawai'i cartographic practices have been "waiting for the right time to be recognized and appreciated," I mean this book could not have been written until cartographic theory evolved from a science of representation to a science of *representational practices*. Cartographers Rob Kitchin and Martin Dodge were the first to propose this theoretical shift, "refocusing attention across the broad spectrum of cartography (practitioners, technicians, historians, critical theorists, map 'users') on understanding *mapping practices*—how maps are (re)made in diverse ways (technically, socially, politically) by people within particular contexts and cultures as solutions to relational problems" (2007, 343, italics original). By focusing on mapping practices instead of mapping products, cartography is now ready to accept representations derived from alternative perceptions of realities. This resituates current popular mapping techniques and technologies as part of an expansive range of mapping possibilities instead of framing anything that differs in a dualistic, and inevitably hierarchical, relationship. The implication of this theoretical shift for Indigenous peoples is tremendous.

It means we can now begin to validate our own cartographic practices instead of having to fit the spatial knowledge those practices maintained into modern technologies that were never created to represent our perceptions of reality. We no longer need to use the master's tools or dwell in the master's house, as Audre Lorde framed the discussion decades ago, albeit in a cartographic sense. Lorde, a feminist lesbian Caribbean American writer, was asked to comment at a paper session on the role of "difference" in the lives of American women at a New York University Institute for the Humanities conference. She was displeased to learn Black women and lesbians were barely included in the discussion.

I make reference to her work here because of two similarities. First, representing Kanaka Hawai'i spatial knowledge with Western cartographic tools, techniques, and technologies that were explicitly created to embody a Western perception of reality implies that any knowledge that falls through the gap does not matter. Second, this paradigm falsely assumes Kanaka Hawai'i have no spatial knowledge or cartographic practices. Embracing and describing our own cartographic reality provides us the opportunity to expand scientific reality, thereby bringing about the genuine change Lorde spoke about when we "take our differences and make them strengths" (1984, 112).

While it is important to use our own cartographic practices to represent our own perceptions of reality, I do not believe we should altogether disavow Western mapping practices. Many Indigenous peoples have successfully used them to help reclaim land titles or define traditional land uses. Cultural geographer Bernard Nietschmann coined the ever-popular statement, "More indigenous territory has been claimed by maps than by guns. This assertion has its corollary: more indigenous territory can be reclaimed and defended by maps than by guns" (1995, 5). At the time Nietschmann was writing about the Miskito Reef Mapping Project, a community-based mapping project that combined geographic and biological knowledge with the experiential knowledge of Miskito sailing captains and affordable mapping technology to defend their traditional sea territories. Nietschmann believed maps made by Indigenous peoples of their own territory challenge their government's ability to erase them from the land/seascape or assert that they have no territory.

He was by no means the first person to engage in this kind of mapping. Four other well-known researchers involved with helping Indigenous communities achieve territorial and land use recognition with modern mapping practices are Terry Tobias, Mac Chapin, Jefferson Fox, and Giacomo Rambaldi. Each of these men have published on their experience in developing tools and methods that help Indigenous communities define their lands and their rights to use their land in different parts of the world.

Terry Tobias is a leading consultant on use-and-occupancy mapping, a methodology that has been used by Indigenous communities in Canada since the 1970s. Tobias has over four decades of field experience designing and conducting various kinds of land use and occupancy research with Indigenous peoples and spent close to eight of those years living and working in aboriginal communities. Tobias wrote the popular *Chief Kerry's Moose: A Guidebook to Land Use and Occupancy Mapping, Research Design and Data Collection*, in 2000, and followed it up in 2009 with *Living Proof: The Essential Data-Collection Guide for Indigenous Use-and-Occupancy Map Surveys*. Both works guide readers through data collection methods intended for maps that will be used as the basis of legal land claims in Canada. Creating maps for this purpose requires the data and methodology to be beyond reproach in order to convince the map reader of the credibility of the research. Tobias outlines a series of research principles and data quality standards needed to achieve this high level of credibility.

Mac Chapin is an anthropologist who has worked with Indigenous peoples for over forty years, largely in Latin America, South America, Africa, and New Guinea. Chapin is cofounder and director of the nongovernmental organization the Center for the Support of Native Lands (Native Lands). This organization assists Indigenous peoples in strengthening their capacity to protect their lands by encouraging open communication with their government, creating collaborative relationships with various conservation and development organizations, and linking them with the technical, legal, and financial resources necessary to carry out their research agendas. Chapin is coauthor of *Indigenous Landscapes: A Study in Ethnocartography*, a narrative account of mapping projects completed in three Central American countries, Honduras, Panama, and Bolivia, which provides a framework for mapping relatively large territories in ethnically complex regions.

Jefferson Fox has been a Fellow/Senior Fellow at the East-West Center for nearly forty years. His research interests include land use and land-cover change in Southeast Asia and the possible cumulative impact of these changes on the region and the global environment. Fox is coauthor of *Mapping Communities: Ethics, Values, Practice*, a book based on a workshop held in Chiang Mai, Thailand, in June 2003. The twenty-three workshop participants, representing eight groups and seven countries, used spatial information technologies (SIT) extensively in their community-based work. The goal of the workshop was to understand the socio-ethical implications of using SIT in community-based mapping projects. The book confirms that SIT is a useful capacity-building resource that promotes decentralization of resource governance in Asia; however, it can come with some unintended results, such as shifting the focus toward exclusive property rights, thereby providing outsiders legal means to gain common property resources.

Giacomo Rambaldi is senior program coordinator at the Technical Center for Agricultural and Rural Cooperation in Wageningen, Netherlands. Rambaldi has been involved with community mapping for nearly forty years. In August 2000 he launched www.iapad.org, a website dedicated to sharing knowledge on community mapping and collaborative spatial information management. He later developed www.ppgis.net, an electronic forum on participatory use of geo-spatial information systems and technologies. He has numerous publications on his research interests, which include visualizing Indigenous spatial knowledge; facilitating peer-to-peer dialogue and conflict management on issues related to the territory; collaborative management of natural resources and protected areas; and participatory spatial planning. He is most noted for his work on developing and promoting participatory 3-D modeling (P3DM), a community-based mapping method integrating GPS and GIS applications that is now used in many parts of the world.

Each of these men helped Indigenous communities make use of modern cartographic or spatial information technologies. The fact that they are all men should not be surprising. When they began their careers four decades ago, the women in cartography did not focus their research on Indigenous communities. However, many women were instrumental in advancing cartographic research on experimental paradigms, "including Olson's (1981) research on design and symbolization (in particular on spectrally encoded bivariate maps), Gilmartin's (1981) research on graduated circles, Caldwell's (1981) research on television maps, and Robinson and Petchenik's (1976) insightful philosophical approach to understanding maps" (McMaster and McMaster 2002, 318).

Nonetheless, since much of the research conducted in Indigenous communities involving modern cartographic or spatial information technologies lacks a substantive female perspective, there is ample opportunity for both women and the discipline in this area. Women relate to and interact with the world differently than men do, and in some Indigenous communities women are not able to speak freely with male researchers. Having a woman researcher working with the women of the Indigenous community provides additional insights to the spatial relationships of the community as a whole. But this point could very easily be the basis of another book. Instead, I would like to shift focus to how Kanaka Hawaiʻi have made use of these modern cartographic or spatial information technologies, specifically through the works of the University of Hawaiʻi at Mānoa Kamakakūokalani Center for Hawaiian Studies and the Office of Hawaiian Affairs.

The most current cartographic works at Kamakakūokalani is the AVA Konohiki project. AVA stands for Ancestral Visions of ʻĀina. *ʻĀina* is the land from which a person eats. A *konohiki* is a land steward who is responsible for maintaining productivity by managing people's consumption of and use relationships with water, land, and agricultural and aquacultural enterprises within a land area known as an *ahupuaʻa*. I discuss this concept further in chapter 3, "Nā Māhele 'Ike Hawai'i (Hawai'i Knowledge Classifications)." The AVA Konohiki project was funded by a grant from the Administration of Native Americans and had two main goals: train students to become modern-day *konohiki*, and make land documents from the Kingdom of Hawai'i, located at the Hawai'i State Archives, accessible online.

Prior to the privatization of land in the 1840s and 1850s, there were 252 known *konohiki*. The land has changed considerably since then, as have the political and economic systems. To train Hawai'i students to become modern-day *konohiki*, the project blends both old and new, taking advantage of both the modern advances in technology and traditional cultural practices. Interested and qualified Hawai'i students take a series of courses at the University of Hawai'i including Hawai'i language, Hawai'i culture, Hawai'i history, land studies, and cartography, and engage in cultural retreats provided by the Edith Kanaka'ole Foundation.

As part of the coursework, students learn how to identify, access, transcribe, and upload land documents, including maps. The project team has not only made these documents accessible online, they have also made them searchable. So, if a Kanaka Hawai'i wants documents about the lands where their family once lived, they can search the website to access scanned images of land records and maps. In some cases land records are transcribed, making them easier to read, as some of the scanned images of handwritten works are difficult to comprehend.

What Kamakakūokalani has accomplished with this project is truly visionary. Many Kanaka Hawai'i take the responsibility for caring for the land and its resources very seriously despite the change in political systems and ravages of the occupied colony's consumer economy. This project teaches Hawai'i students how to use the system and its resources to carry out their responsibility as *konohiki* of their *kulāiwi* for the benefit of everyone. While this project focuses on blending traditional cultural practices with modern technology in order to prepare the next generation as cultural practitioners, the Office of Hawaiian Affairs has incorporated this archival data into a geographic information system (GIS), making real world locations searchable.

The Office of Hawaiian Affairs has maintained a GIS for nearly two decades, but it took the ingenuity of the current staff to create a geographic database, named Kīpuka, that helps answer questions about land use. A *kīpuka* is a variation or change in form, such as an oasis within a lava bed. The Office of Hawaiian Affairs is using Kīpuka as a repository of Hawaiʻi geographic knowledge, blending culture with land and history. Understanding why cultural sites exist where they do provides a glimpse into the precontact Kanaka Hawaiʻi land planning process. For example, the placement of certain *hālau*, Hawaiʻi place of learning, depended on the resources needed for the operation of the structure. A *hālau* for making *kapa*, bark cloth, needs to be near freshwater. A *hālau* for *waʻa*, canoe, needs to be near the coast.

The Kīpuka database is a virtual *moʻokūʻauhau*, genealogy, of land tenure in Hawaiʻi, including the ceded lands. Ceded lands are the 1.8 million acres of land that belonged to the Kingdom of Hawaiʻi prior to the overthrow by annexationists on January 17, 1893. Initially, all these lands were crown lands, as established in the Māhele of 1848 by Kauikeaouli (Kamehameha III). Kauikeaouli eventually granted some of his lands to the Kingdom of Hawaiʻi for the purpose of running the government. Both government and crown lands were ceded to the United States upon annexation in 1898, pursuant to a joint resolution known as the Newlands Resolution. This was a travesty of international justice.

Twenty nations including the United States held treaties with the Kingdom of Hawaiʻi, recognizing it as a sovereign and independent nation among the world community of nations. Acquiring another nation requires a treaty, not an act of Congress. This act of treachery worked because the annexationists wanted tariff-free sugar exports and the United States wanted to control the Pacific Ocean for the remainder of the Spanish-American War in the Philippines.

When Hawaiʻi became a state on August 21, 1959, the United States returned the ceded lands and authorized the State of Hawaiʻi to hold those lands in a trust, using the revenues for five purposes: support of public education, betterment of the condition of Kanaka Hawaiʻi, development of farm and home ownership, public improvements, and public use. The 1978 Hawaiʻi State Constitutional Convention created the Office of Hawaiian Affairs, a semiautonomous entity, as a public land trust. Its mandate is to better the conditions of both Kanaka Hawaiʻi and the community in general, with a pro rata share of revenues from the ceded lands held by the State of Hawaiʻi.

The Office of Hawaiian Affairs has a record of the current condition of the ceded lands according to the land inventory database maintained by the State of Hawaiʻi. Many people have attempted to trace land records back to the Māhele of 1848, with mixed results. The Kīpuka project takes a different perspective. It begins with the original land divisions that created the crown, government, and private lands. From this original inventory, they remove those parcels that were sold or transferred, legally or not so legally, attaching the relevant documents to their geographic information system. This repository of knowledge not only provides a better picture of the overall genealogy of ceded land tenure, it also provides easy access to information about all of Hawai'i's land, culture, and history, thereby giving all people the opportunity to forge new relationships with those lands that are most important to them (Office of Hawaiian Affairs 2013).

Both Kamakakūokalani and the Office of Hawaiian Affairs have found innovative ways of utilizing modern spatial information technologies to provide a better picture of the current state of land use and occupancy in Hawai'i. However, they were not the first Kanaka Hawai'i to use modern technologies. According to Hawai'i political geographer Kamana Beamer, the *ali'i*, Hawai'i political leaders, modernized the government "through the modification of existing indigenous structure and through Hawaiianizing Euro-American structures to suit their own needs" (2008, v). Beamer challenges arguments that the Kingdom of Hawai'i was a colonial institution by situating and interpreting archival documents, such as maps, in their historical context.

Maps are extremely important to the colonial process. Benedict Anderson identifies maps as one of three institutions that "profoundly shaped the way the colonial state imagined its dominion" (2006, 163–164). The other two institutions are the census and the museum, both of which, I agree, are established organizations of information. However, I take issue with Anderson's use of the term "maps" as an institution. Maps are no more an institution than a census spreadsheet or a museum sculpture. They are momentary glimpses of spatial reality framed by the author, which, in this case, is the colonizing power.

Cartography is the institution that profoundly shaped the way the world was viewed by Europeans, and surveyors implemented that worldview by accurately measuring, recording, and ordering territory. The exacting nature of European cartography emerged during the Renaissance and was especially useful in the age of exploration. Technological advances and philosophical movements in the seventeenth and eighteenth centuries made "scientific" cartography indispensable for imperial powers to assert "control over familiar and domestic peoples and territory as well as more distant and alien places" (Akerman 2009, 2). Although the techniques used to maintain control differ from those used to attain control, colonial cartographic projects around the world were distinctive because of the way they framed nature, imposed a preferred reality, and legitimized the conquest of territory.

Timothy Mitchell describes the colonizing process in Egypt as an "enframing" of imposed internal order through a panopticon design of the division or partitioning of space, isolation of the population, and implementation of systematic surveillance (1988). Thongchai Winichakul expresses how the Thai identity and nationhood evolved in conjunction with its geographic consciousness. Where once the concept of territory precluded the delineation of fixed boundaries, allowing for overlapping sovereignties or unclaimed buffered corridors, then came the modern geo-body, arbitrarily and artificially created by European cartographic confrontations, that displaced cultural discourse (1994). Matthew Edney examines the overlap between mapmaking and empire-building in India, arguing that maps created a system of knowledge and became the symbol of property and domination (1997). Ian Barrow extends Edney's work and reflects on how British colonial maps of India enabled imperial authority to produce its own reality and legitimize its power (2003). Giselle Byrnes documents the role British surveyors played in preparing the New Zealand landscape both physically and psychologically for British settlement. Byrnes focuses on how the land was transformed through the imposition of British place names and town grids as "deliberate and provocative statements of power" (2001, 80).

The common thread holding each of these authors' works together is Indigenous erasure. Most maps produced by cartographers and surveyors during the colonial era marginalized Indigenous peoples, silencing Indigenous spatial knowledge systems, which led to, in many cases, irrevocably severing Indigenous relationships with their cultural landscapes. However, Beamer argues that maps made during the colonial era by the Kingdom of Hawaiʻi maintained Hawaiʻi place names and preserved Hawaiʻi political land divisions, which in effect maintained existing resource relationships.

Although Beamer's argument is based on a Hawaiʻi perspective of government, I believe using spatial technologies in cross-cultural situations severely limits the spatial realities of the "other" culture, much like playing on a piano a piece of music composed specifically for a violin. Something is lost in the performance. This loss, whether intentional or not, is just as devastating as the loss experienced on the battlefield. Loss of honor, dignity, and identity leads to generational grief and defiance. Hawai'i political scientist Sydney Iaukea expresses her frustration toward the cartographic endeavors of the Kingdom of Hawai'i, stating, "Even the early native maps demanded erasures in the landscape. There was simply no space, no language, no preserving the personal relationships and the social persona of the community as a whole" (2012, 26). This is exactly the kind of problem the cross-cultural use of cartographic technology can create. Although the Kingdom of Hawai'i never intended to supplant Kanaka Hawai'i cartographic practices, the disconnect it created between the embodied landscape and cultural memory has lasted generations.

This sentiment about maps and the emotion it invokes with Indigenous peoples worldwide is real. I have experienced it on many occasions and can empathize. My first reaction was righteous rage that I still tap in to from time to time. It took a few years to learn how to focus on the issue instead of my reaction. Now, I am at a point that I want to do something positive with all that rage. Releasing my anger allowed me to see pathways filled with creation, connection, and compassion. It was when I chose to walk this path that I realized how I could best contribute to the ongoing process of acknowledging the existence of Indigenous cartographies worldwide. This book is my contribution to that endeavor.

I recall the moment I decided to walk this path, at the 2004 International Forum on Indigenous Mapping in Vancouver, Canada. Alvin Warren invited me to attend the forum as the closing plenary speaker, an honor that I ultimately shared with my colleague, Indigenous geographer Jay Johnson. I had no idea what I was going to say. I decided not to prepare a presentation ahead of time, opting instead to listen to what Indigenous peoples were doing with mapping before deciding how to close the gathering. Although Jay had prepared a speech, he decided to rework it based on our observations. What was absolutely clear to both of us was that there was little to no acknowledgment from any of the presenters we saw that they had their own cartographic traditions.

We had to work together all night long and got absolutely no sleep the night before our presentation. We decided he would speak first and theoretically ground our insights, while I summarized the presentations and added a few comments about how these cartographic engagements could be transformed from Indigenous people using mapping techniques and technologies to Indigenous mapping. I have since become confident enough to call it Indigenous cartography, noting that mapping is a cartographic process and not vice versa. The title of our presentation was "Facing Future," which we later published as an online article with another colleague from the University of Hawaiʻi, Albertus "Monti" Pramono.

Kanaka Hawaiʻi cartography is similar to Western cartography in that it is an intimately interactive, dynamic, and contextual process-oriented activity, the main goal of which is to provide a shorthand communication system of understanding spatial phenomenon. It is distinctive in that Kanaka Hawaiʻi cartography places emphasis on multisensual cognitive abilities, multidimensional symbolic interrelationships, and performance as a primary mode of communication. This difference in emphasis is directly related to how Kanaka Hawaiʻi philosophical underpinnings affected its cartographic development, which is described in part 1 on Kanaka Hawaiʻi cartographic foundations.

Every major undertaking requires a certain degree of preparation. To understand the practices of another culture, it is best to immerse yourself in that culture. Committing oneself to learning the language, working/walking/wading the land/waters, eating the foods, and fitting in with the social structures is not easy. It can be frustratingly difficult or transformative, but it is the best way to truly understand another culture's nuances. Short of providing that kind of cultural experience, I have chosen to begin your journey of understanding Kanaka Hawaiʻi cartographic practices with a description of the basics of Kanaka Hawaiʻi cartographic philosophy, followed by a presentation of Kanaka Hawaiʻi cartographic knowledge.

Kanaka Hawaiʻi cartographic philosophy is a discussion of how the sensual and cognitive elements of Kanaka Hawaiʻi spatial perception inform spatial knowledge acquisition, symbolization, and transmission. Kanaka Hawaiʻi cartographic knowledge builds on this philosophical underpinning, presenting how Kanaka Hawaiʻi spatial/temporal order and classification informs cartographic knowledge production. By the time you complete part 1, you will be ready to understand the Kanaka Hawaiʻi cartographic practices described in part 2.

Part 2 presents three Kanaka Hawaiʻi cartographic practices: *ka hoʻokele*, *ka haku ʻana*, and *ka hula*. Each practice takes years of preparation, mentoring, and training before a person is ready to navigate a long-distance voyage, compose any oratory, or choreograph any dance. These particular cartographic practices also integrate other cultural practices. For example, one does not just jump into a canoe and sail from Hawaiʻi to Tahiti and back. The planning and leadership necessary to conduct a large-scale voyage is quite phenomenal, as is the degree of knowledge necessary to compose works of oratory that span generations or the amount of physical training necessary to grace the stage at a *hula* performance.

As I believe new discoveries are possible when we take something familiar and see it from a new perspective, it is my hope that my fellow Hawaiʻi scholars will learn more about their own areas of expertise through my composition, arrangement, and understanding of Kanaka Hawaiʻi cartographic practices. Understanding the world through a Kanaka Hawaiʻi cartographic lens is what I offer. As you begin this journey, allow me to provide the first conceptual shift. I have paired seemingly disparate concepts with an up slash, such as space/time, mind/body, and observer/participant. From a Kanaka Hawaiʻi cartographic perspective, these concepts are not separate but distinctive. Please know I am not trying to coin new concepts. I believe these pairings help to distinguish Kanaka Hawaiʻi cartographic reality.

PART 1

Nā Kahua Hawaiʻi

(Hawaiʻi Foundations)

Many Kanaka Hawaiʻi practitioners believe knowledge is linked to both its genesis and its progeny, similar to the naturally occurring spirals found in the ocean, atmosphere, among the flora and fauna, and even in deep space. As the tendril of a fern unfurls, embedded within it is the genetic makeup from generations past, yet also within it is the ability to adapt, allowing the next generation to endure the current conditions and thrive. These practitioners recognize that similar patterns exist in deep space, the atmosphere, and on earth, and maintain a knowledge framework based on metaphoric logic connecting seemingly dissimilar entities.

This belief and framework is reflected in practitioners' methods for recording and sharing their perspectives of scientific knowledge about the world around them. Much of Kanaka Hawaiʻi cartographic knowledge was demonstratively passed from one generation to the next, ensuring knowledge was embodied in rhythmic movement, tonal qualities, and mnemonic composition. Much like the unfurling fern, these movements, tones, and compositions ingeniously embody a metaphoric logic that could adapt to the present day without losing the meaning of the past. Weaving Kanaka Hawaiʻi perspectives of scientific knowledge into storied accounts, with concepts becoming characters, creates multilevel knowledge access such that children remember them, adults maintain them, and masters decipher them. Kanaka Hawaiʻi practitioners incorporated these levels of understanding and modes of communication into their cartographic practices.

The chapters in part 1, "*Nā Kahua Hawaiʻi*," prepare you to observe the world through a Kanaka Hawaiʻi cartographic consciousness. In this text, cartographic consciousness is the ability to examine a situation using spatial processes, such as determining the most efficient way of completing daily errands in town or picking flowers to make a *lei*. Kanaka Hawaiʻi cartographic practices incorporate these processes in a manner that is uniquely appropriate to their evolving traditions. Cartographic consciousness comes first. A person must be aware that their ability to observe the world is framed by a specific lens. Cartographers observe the world differently from political scientists or economists. These frames or lenses are also culturally tuned. Cartographic process comes next. Once the world is understood in a particular way, a particular series of actions is performed in order to bring about some kind of result or product. A person qualified in medical trauma will act differently from one trained in criminal science when they happen upon the same accident at the same time. Key cartographic processes include spatial/temporal knowledge acquisition, representation, and transmission. These processes or series of actions are also culturally influenced. Cartographic practices engage those processes according to a particular relational issue they clarify, refining them until they become a habitual or customary expression of spatial/temporal knowledge.

Understanding the distinctiveness of Kanaka Hawaiʻi cartographic practices begins with a description of a Kanaka Hawaiʻi cartographic philosophy that integrates the physical realm with its social and spiritual counterparts. Much of the information in part 1 was compiled from the works of S. M. Kamakau, Kepelino Keauokalani, John Papa Ii, and David Malo (Kamakau 1964, 1976, 1991, 1992; Keauokalani 1932; Ii 1995; Malo 1971). These *kāne*, men, published in the mid- to late nineteenth century on Hawaiʻi culture. Many Hawaiʻi scholars have returned to their original Hawaiʻi language texts to tease out for themselves the knowledge shared by these authors.

As a counterbalance to these *kāne* are the works of *wāhine*, women, including Mary Kawena Pukui, Rubellite Kawena Kinney Johnson, Pualani Kanakaʻole Kanahele, Kekuhi Kealiʻikanakaʻoleohaililani, and Kalei Nuʻuhiwa (formerly Kalei Tsuha), among others (Johnson 1993, 2000, 2012, n.d.; P. Kanahele 2005, 2014; Kanahele 2001; Kanahele et al. 2009; P. K. Kanahele 2011; Kealiʻikanakaʻoleohaililani 2013a, 2013b, 2013c, 2013d; Pukui 1983; Pukui and Elbert 2003; Pukui, Elbert, and Moʻokini 1974b; Pukui, Haertig, and Lee 1972; Pukui and Korn 1979; Tsuha 2008). Their rigor and passion for understanding the cultural mores and traditions of their ancestors is surpassed only by their ability to clearly and effectively share that learning with others.

CHAPTER ONE

Nā Kumu Manaʻo Hawaiʻi

(Hawaiʻi Theoretical Sources)

When first asked whether Kanaka Maoli had a word for "cartography," I could not respond—not because I did not know, but because I had no idea how cartography was practiced prior to the arrival of Cook in 1778. I knew Kanaka Maoli shared knowledge orally but I had yet to learn how they communicated spatial knowledge, the purview of cartography. If mapping is indeed a cultural and "cognitive universal" (Blaut et al. 2003), then effectively and efficiently communicating spatial knowledge was necessary for Hawaiʻi society to flourish. But what exactly is spatial knowledge? According to human geographer Tuan (2001), spatial knowledge should not be confused with spatial skill. Spatial skill precedes spatial knowledge and has to do with our ability to perform our daily routine. It is what our bodies are capable of doing and, in that sense, is similar to agility, dexterity, and mobility. Spatial knowledge enhances and extends our spatial ability, whether it has to do with athletic prowess, intellectual acumen, or cultural achievement. It allows us to confidently increase our range of mobility beyond our local physical geography.

For example, I am a confident driver in Hawaiʻi. I can get from one place to another without a road map because I know where the mountains are in relation to the ocean. However, when I drive in the continental United States, I am less confident in getting from one place to another without a road map. Nonetheless, after a daylong field trip in a car full of cartographers, I was the only one capable of navigating back to our hotel without using the instructions or referring to a map. I have always known I can get myself out of a place by retracing my steps, but this was the farthest I had traveled, and I was pleasantly surprised. The real surprise came when I could not remember what turn to make at an intersection because changes in daylight obscured the background scenery. Yet, turning left just felt wrong. My body wanted to turn right, which was the correct choice according to those referring to the map. I wonder, though, did I know the right turn was the correct choice because I trusted my body's spatial memory, or was it a product of my Kanaka Hawai'i cartographic conditioning?

My first attempt at defining Kanaka Hawai'i cartography was to focus on key cartographic processes of acquiring, symbolizing, and transmitting spatial knowledge. However, when Kitchin and Dodge defined cartography as "the pursuit of representational solutions (not necessarily pictorial) to solve relational, spatial problems" (2007, 343), I realized that defining Kanaka Hawai'i geographic and cartographic fluency requires an understanding of Hawai'i philosophy. I first learned of the distinctiveness of Hawai'i philosophy from Hawai'i educator Manulani Aluli Meyer in 1998. I have spent more than a decade swimming in the ocean of knowledge she describes in her dissertation, which she published, unedited, in her book, *Ho'oulu—Our Time of Becoming: Hawaiian Epistemology and Early Writings*. Yet, I still find myself teasing out new ideas from its depths. Meyer reviewed historic and contemporary literature and interviewed twenty *kūpuna*, Kanaka Hawai'i elders, to answer the question, "What is Hawaiian epistemology?"

From her work, I learned that many Kanaka Hawai'i consider knowledge as experiential, sensory, embodied, and profoundly linked to place. This is very similar to Shawnee philosopher Thomas Norton-Smith's interpretation of American Indian philosophy, described in his book, *A Dance of Person and Place*, where he examines four common recurring themes: "two world-ordering principles, relatedness and circularity, the expansive conception of persons, and the semantic potency of performance" (2010, 1). Like Norton-Smith, I must include a couple of caveats about expressing a Kanaka Hawai'i cartographic philosophy.

The first is that there is no such thing as *the* Kanaka Hawai'i cartographic philosophy. While many people like to bundle all of the Islands of Hawai'i into a single sociocultural entity, many other people argue this is not an accurate perception; it would not surprise me if this book inspires some to express that other Kanaka Hawai'i cartographic philosophies exist. The second is that Kanaka Hawai'i do not separate their understanding of the world into the same silos of knowledge found in Western universities. In fact, the Kanaka Hawai'i cartographic practices identified and described in this text can be used by a Hawai'i historian, literary scholar, political scientist, or educator to further understanding in their own disciplinary silos of knowledge. This mutability emphasizes the existence of a pluriverse of knowledge systems that, once embraced, will lead to a much-needed intellectual transformation.

So, if there are Kanaka Hawai'i who believe knowledge is spatially relevant, then there must be culturally specific philosophical paradigms that greatly affect what they consider spatial knowledge to be, from how they acquire it and the ways in which they encode and share knowledge with one another to how they pursue representational solutions in order to solve relational, spatial problems. Every culture creates distinct protocols and practices to convey its spatial representations. The most commonly known form of spatial representation is a map-like product, whether it is on paper, the Internet, or a Global Positioning System. However, maps are not just a diagrammatic representation of an area.

Maps are symbolic abstractions of experienced spatial phenomena that assist in communicating solutions to problems of a relational or spatial nature. They consist of myriad pieces of information compiled in such a way that they reflect each culture's perception of reality. Maps are a shorthand representation of otherwise perplexing or confusing spatial phenomena or spatial processes presented in manageable portions through sociocultural lenses. Different cultures impose very different lenses on their experience of the world and don't necessarily experience the same world with different labels. For example, a mountain may mean quite different things to different societies. In Hawai'i, one need only look into the struggle between some Kanaka Hawai'i and astronomers over the building of another telescope on Maunakea to realize this difference.

Many Kanaka Hawai'i relate with Maunakea as a *wahi pana*, a celebrated place, a place that holds significance for the people who live there. These places are treated with great respect, honor, and reverence. More often than not, *wahi pana* is translated as "sacred place," which unfortunately makes some people perceive it as a religious object. It is true that some Kanaka Hawai'i relate to Maunakea as the realm of the *akua* and tell stories of the Poli'ahu when snow dusts the top of Maunakea; let me take a moment to explain my understanding of *akua*. Although many people have casually referred to *akua* as god, the Kanaka Hawai'i understanding of god is not the same as the Western conceptualization. In my understanding, *akua* is a particular type of entity with specific responsibilities that we associate with those natural processes necessary for life as we know it to exist. As such I consider them divine entities with whom we can interact. The top of Maunakea is where many life-giving natural processes can be observed and interacted with, from an intimately personal scale to a rather ritualistic enterprise. However, this is not the only reason Kanaka Hawai'i believe Maunakea is a *wahi pana*; another reason Maunakea is revered and considered sacred is because it is capable of attracting, harnessing, collecting, and storing large quantities of freshwater—the most highly prized natural element.

Astronomers see a mountain peak formed by a shield volcano situated above the clouds with very little light noise or wind disturbance to hinder their ability to peer into the darkness of space. In order to fund a project using newer technology that will allow them to see deeper into space, they appeal to investors by telling them that US dominance in astronomical discovery is slipping. The investors hire project planners who know their way around Hawai'i State Board of Land and Natural Resources conservation district rules. The project planners contend the building of a new telescope cannot further desecrate a mountaintop already violated by previous projects. This is in direct opposition to a 2005 court-ordered Environmental Impact Statement that concluded the cumulative impact of thirty years of astronomy activity caused "significant, substantial and adverse" harm. Nonetheless, the University of Hawai'i Institute for Astronomy complies with the State Third Circuit Court's finding to create a Comprehensive Management Plan for Maunakea, which is approved by the State of Hawai'i Board on Land and Natural Resources, paving the way for an unlimited number of telescopes to be constructed in the 525-acre Astronomy Precinct at the summit of Maunakea.

Many Kanaka Hawai'i continue to oppose the construction of telescopes, not because they are opposed to science or astronomy in particular, but because the location violates one of their cultural imperatives—to protect sacred ground. I would say the proponents of this new telescope do not get the point being made, but part of me believes they absolutely understand what they are doing, and they are choosing to use the laws their culture created to privilege their values of "progress" and justify what they want to do. Indigenous peoples around the world continue to witness how this kind of scientific arrogance often leads to multiple tiers of ethical misconduct and, in this particular situation, screams of the colonial legacy Kanaka Hawai'i have faced for centuries.

Mana as a Kanaka Hawai'i Ontology

Accepting a mountain as anything but an object, a commodity to be bought, sold, and used as mankind sees fit, requires an ontological shift, a different theory of existence. The popular version of reality is that man is in charge because of his capacity to reason, and therefore everything in the world is at his disposal to sustain his life. A Hawai'i cartographic ontological understanding of reality is that everything is imbued with varying degrees of *mana*. The most common understanding of *mana* is power, strength, might. It is also understood to be supernatural power, the kind of power attributed to *akua*. When Christians came, the word adopted new meanings, including religion, divine power, and worship. Various Hawai'i language dictionaries also include the following explanations for *mana*: spirit, energy of character, glory, majesty, intelligence, a line projecting from another line like a branch of a tree or of a stream, a variant or version as of a story, a limb of the human body and the stage of growth in which a fetus forms limbs, the stage of growth of a fish in which colors appear, the name of a specific house on a *luakini heiau* (a place of worship of the highest order), a variety of *kalo* (taro) used as medicine that propagates by branching from the top of the corm, a native fern with several large subdivided fronds, and a particular type of fishhook used to catch eels (Andrews 1865; Andrews et al. 1922; Pukui and Elbert 2003). Hawai'i language is efficient. Think carefully on each of the explanations and you will begin to understand the connection between them.

To me, *mana* is an intangible life force, and many Kanaka Hawai'i recognize that everything has a life force. Everything has a life cycle, a sequential progression of generation, regeneration, and degeneration. Norton-Smith argues that this circularity is a world-ordering principle, where the focus is on space, place, and nature as opposed to time, events, and history (2010, 119). This focus allows Kanaka Hawai'i to recognize and construct a world where inanimate objects such as rocks have *mana* because they are birthed from the earth and they eventually wear away and return to the earth. Yet, their time on the earth is not without purpose. Rocks help things grow or provide places for other entities to live. For these Kanaka Hawai'i, this means the rock has some form of intelligence, perhaps not cognitive intelligence, seemingly reserved for human people, but certainly an innate intelligence, which is one of the many explanations of *mana*.

ʻOhana as a Kanaka Hawaiʻi Epistemology

Many Kanaka Hawaiʻi honor these distinctive intelligences by constructing an epistemological framework based on *ʻohana*, or family, relative, kin group, related, a group of kindred individuals, lineage, race, tribe, those who dwell together (Andrews 1865; Andrews et al. 1922; Pukui and Elbert 2003). They maintain a different understanding of *ʻohana* than what we now know as a "nuclear family." According to anthropologist E. S. Craighill Handy and Hawaiʻi scholar Mary Kawena Pukui, *ʻohana* "extends far beyond the immediate biological family. . . . [It] must be thought of against the background of the whole community of kith and kin, including in-law, and adoptive categories" (2010, 40). Familial relationships also extend beyond the interpersonal or social to include "all nature, in its totality, and all its parts separately apprehended and sensed as personal" (118). These Kanaka Hawaiʻi recognize dynamic relationships between all life forces, including the elements and processes of nature as familial relationships. This intimate and sensual participation with nature as family is a cultural conditioning or tuning in to the natural world and is oftentimes mistaken for supernatural, extrasensory, or extra-ordinary abilities.

Handy and Pukui do not believe any of these terms correctly describe the Kanaka Hawaiʻi kinship with the natural world. It is not "extrasensory" because it is both "of the senses" and "not of the senses" (2010, 117). Cajete says these abilities are a naturally conditioned response to nature. A person who tunes in experiences the world with greater sensual acuity and can be trained to detect subtle shifts in nature with heightened and discriminating sensitivity (2000, 20). Basso prefers to call it the "sensing of place," because it is a continual multisensory "way of engaging one's surroundings and finding them significant" (1996, 143).

The *ʻohana* epistemological framework places humankind as part of a natural knowledge network. Everyone in this world—fish people, bird people, rock people, cloud people, plant people—all do and act according to their genealogical intellect. Norton-Smith explains this as an expansive conception of persons, where moral agency occupies a core principle (2010,

12). We may understand each people's habits and their life cycles, but we do not always understand their intellectual ability. Many Kanaka Hawai'i have such great respect for the intelligence of nature that when they adapt to the natural environment they mimic the natural processes they observe.

For example, wetland taro cultivation diverts water from the main stream without destroying the natural ecosystem of the stream. Water flows through *lo'i kalo*, taro ponds, and back into the main stream enriched with organic material, similar to the natural processes of a stream ecosystem. This enriched freshwater drains into the ocean, where Kanaka Hawai'i observe reef fish thriving. So, they erect fishponds using sluice gates that allow small fish to pass until they reach a certain size. The fishpond provides small fish a safe place to roam freely and feed on the nutrient-enriched waters from the irrigation system. When they become too large to pass through the sluice gates it is time to harvest.

The harvest is necessary to keep the numbers reasonable for continued growth. Any fish small enough to escape harvest through the sluice gates are considered too small to eat. Inevitably the fish return to the safety of the fishpond until they give themselves to the people who feed and protect them. Yes, many Kanaka Hawai'i still believe the fish give themselves as nourishment, because the fish are intelligent enough to escape capture if they truly want to. The choice to remain in the fishpond is perceived as reciprocity, giving back to those with whom intimate bonds are formed. Many Kanaka Hawai'i recognize these "cycles of life" as part of the resiliency of landscapes or, in this case, reef-scape. Inserting themselves and their needs into this natural cycle is done in a manner that maintains sustainability for generations to come, but more importantly, it respects nature as an intelligent being.

Many Kanaka Hawai'i recognize natural elements and processes as divine entities that are oftentimes pigeonholed or typecast into the role of deity. I assure you, these divine entities are so much more than that, as Handy and Pukui so eloquently share:

> It is hard for the modern intellectually regid [*sic*] and extroverted mind to sense the subjective relationship of genuine Hawaiians to Nature, visible and invisible. But without in some degree *sensing the feeling* that underlies this quality of consciousness in those who live intimately in a condition of primary awareness and sensitivity on the plane of subjective identification with Nature, coupled with perceptions and concepts arising therefrom—without

> some comprehension of this quality of spontaneous *being-one-with-natural-phenomenon which are persons, not things*, it is impossible for an alien (be he foreigner or city-hardened native) to understand a true country-Hawaiian's sense of dependence and obligation, his "values," his discrimination of the real, the good, the beautiful and the true, his feeling of organic and spiritual identification with the ʻaina (home-land) and ʻohana (kin). (2010, 28, italics original)

Maintaining relationships with elements and processes of nature requires continually practicing protocols and compliance. I do not mean to make this sound laborious, because it is not. Many Kanaka Hawaiʻi approach this kind of relationship much like they do any other relationship. There are rules of etiquette for just about every relationship, from business to dating, for men and women, even for children. When two people meet for the first time, there are rules of etiquette people generally follow that are culturally conditioned. In Hawaiʻi, most times, we greet each other with a hug and a kiss. In Japan, people greet each other by bowing, ranging from a small nod to a deeper bend at the waist depending on social status. In most parts of the United States, people who meet for the first time greet each other with a handshake. Breaking these rules or not following them comes with consequences, small or large, whether it is a snide comment or the loss of a business contract. Following the rules sometime comes with compromises. Many people consider it rude when others are late for social engagements. Yet others believe arriving "fashionably late" is perfectly okay. Depending on the situation and preference of time etiquette, there may need to be some acquiescence and compromise.

The point is, relationships come with protocols or rules of etiquette or interaction. For many Kanaka Hawaiʻi, maintaining relationships with divine entities means continually practicing appropriate protocols. Before entering a person's home, calling ahead, whether a few days or just ringing the doorbell, is a considerate gesture. The same is true about entering a forest. Some Kanaka Hawaiʻi will *oli,* chant, prior to entering a forest, a Hawaiʻi form of calling ahead. The *oli* will vary depending on the reason for entering the forest, whether it is to gather plants for adornment or medicine, seek answers, or just leisurely share in the sensual awakening of the scents, sounds, and sensations. The more you seek from the divine entity, the more the need to give up something of value, a sacrifice. In some cases these sacrifices were as simple as a daily *pule,* prayer. In others, *makana,* gifts, were given in exchange for consideration of requests.

There are numerous stories of *kūpuna* who interacted with nature and natural elements regularly. In her book, *Change We Must*, Nana Veary, Hawai'i spiritual scholar, recounts a story of her mother's afternoon luncheon. She set a beautiful table outside under a tent, but the sky grew darker, threatening to rain. Her mother went to the corner of the yard and addressed the dark clouds, humbly saying that she was about to have a luncheon and did not want to be rude with her request to have the rain go to another valley until after the luncheon was over. The skies cleared up as the luncheon began and poured soon after it ended. Veary clearly states, "Hawaiians had the knowledge to work with the elements but they had to be genuinely humble.... When my mother asked the rain to leave our valley, she held the attitude of complete reverence" (1989, 32).

Hawai'i geographer Carlos Andrade asserts these kinds of interactions with the *'āina*, land from which a person eats, are prerequisites of what he identifies as a Hawaiian geography. He eloquently likens the mystery associated with these kinds of interactions as far from a delusional symptom of madness but more akin to the modern high-tech discovery of "conversations" elephants have with each other over vast distances using low frequencies. Andrade's Hawaiian geography supports the idea that the *'āina* talked to him, counseled him, and directed his actions as he restored the place where his ancestors invested themselves, their essence, into *lo'i kalo* (2014).

Kinolau and *Kaona* as Kanaka Hawai'i Methodological Processes

If some Kanaka Hawai'i have the knowledge to work with natural elements, how is that knowledge structured, how is it organized, how is it maintained? Kanaka Hawai'i developed methodological processes known as *kinolau*, many bodies, and *kaona*, multiple meanings. *Kinolau* is a process of grouping natural entities together based on similar characteristics, roles, and responsibilities, whether it is of color, form, pattern, or behavior. For example, Lono, the entity associated with agricultural fertility, peace, and recreation, has many *kinolau*. Seasonally, Lono occupies the islands during *ho'oilo*, wet season, when the rain clouds and winter storms bring much-needed water to the dry, leeward coasts of the islands, helping to keep lands fertile. In the atmosphere, Lono is clusters of dark clouds, thunder, lightning, rain, wind, partial rainbows, whirlwinds, and waterspouts. As landforms, Lono is the gushing springs on the mountains, rocks washed down ravines, and red-stained mountain streams. Lono is also the *ʻuala*, sweet potato, *kukui*, candlenut, *puaʻa hiwa*, black pig, and the *ʻanae*, mullet fish. The abundance of *kinolau* associated with Lono is an indication of his importance in daily life. All the major entities have multiple *kinolau*, reflecting both the observational acuity of Kanaka Hawaiʻi and their ability to systematically store knowledge (Handy and Handy 1972).

Handy and Pukui describe the *kinolau* methodological process as "logic by analogy," whereby observed natural occurrences resemble celebrated characteristics of ancestral beings from as far back as the primordial movements of natural entities to those *kūpuna* elevated to the rank of *akua* because of the *mana* they possessed to accomplish deeds deemed important (2010, 122–123). Cajete refers to it as the "metaphoric mind," the elder brother of the rational mind, where creativity, perception, physical senses, and intuition manifest in "abstract symbols, visual/spatial reasoning, sound, kinesthetic expression and various forms of ecological and integrative thinking" (2000, 30). Norton-Smith explains this "relatedness as a world-ordering principle," wherein these relationships are one way of ordering sense experiences (2010, 10). For the Kanaka Hawaiʻi practitioner, everything is related at all times, and the knowledge structure they maintain embeds multilayered meanings reflecting this metaphoric logic.

Kaona is a methodological process similar to *kinolau* in that it recognizes a multiplicity of meanings embedded in Hawaiʻi linguistic structures, and its expression has been expounded on by several Hawaiʻi scholars in various disciplines. Noenoe Silva, Hawaiʻi political scientist, recognizes how it was used as political expressions of sovereign solidarity (2004). Hawaiʻi historian Noelani Arista believes it questions the validity of monoperspectival Euro-American interpretations of contact, colonization, and resistance (2010). Brandy Nalani McDougall, Hawaiʻi literary scholar, sees it as a way to connect with Hawaiʻi ancestors and an affirmation of Hawaiʻi aesthetic sovereignty (2014). Stephanie Nohelani Teves, Hawaiʻi ethnic studies scholar, emphasizes how it was used as a form of resistance, along with *mele kūʻē*, Hawaiʻi poetic oratory of resistance (2015).

Once meant for specific audiences or perhaps employed to exclude those less-deserving listeners, *kaona* is now widely recognized as having a multiplicity of meanings layered into oral, and now literary, compositions providing knowledgeable listeners a depth of understanding. Alternative explanations can be metaphorical, both figuratively and literally, allegorical or symbolic in origin, all of which implies *kaona* is a complex interconnected system of signification. Many Hawaiʻi *ʻōlelo noʻeau*, proverbs, are scarcely understood without interpretation, for example:

> *Hoʻokāhi e pōʻino, pau pu* [*sic*] *i ka pōʻino*. One meets misfortune, all meet misfortune. Said of those who are important to the community—when misfortune befalls one, it is a misfortune for all. The fall of an able war leader is a disaster to his followers just as the fall of a good warrior is a disaster to the leader. Every member of the group is important. (Pukui 1983, 114)

The literal explanation given is deepened by the commentary that follows. Sometimes these proverbs were derived from stories, for example: "*Kau pō Kāneiahuea*. All night long rode Kāneiahuea. Said of one who wastes time in useless effort. From the story of a man who started out from the inlet of Kāneiahuea, Kona, one night. Because he was unfamiliar with the place, he went back and forth all night without finding an outlet to the open sea" (Pukui 1983, 176). Yet others still require the listener or reader to understand Hawaiʻi cultural mores, for example, "*He kumu kukui i heʻe ka pīlali*. A kukui tree oozing with gum. A prosperous person" (Pukui 1983, 79). Unless you understand how the oozing of a *kukui* tree is connected to a prosperous person, the *kaona* of this *ōlelo noʻeau* will not be grasped.

According to scholar of Hawaiʻi language and anthropology Puakea Nogelmeier, it may indeed be nearly impossible to truly understand the *kaona* of any oral or literary composition, as "Kawena Pukui says nobody can understand kaona." So, he believes *kaona* is "what inspired the poet to write . . . [and can] . . . never be known outside the circle of the poet and those that are real close. It is the seed that made the piece come about" (2010). Whether either *kaona* or *kinolau* are completely understood, both these methodologies indicate that Kanaka Hawaiʻi valued metaphoric multiplicity.

Papakū Makawalu as a Kanaka Hawaiʻi Methodology

Recently, *kumu hula*, Hawaiʻi dance teacher, and Hawaiʻi scholar Pualani Kanakaʻole Kanahele of the Edith Kanakaʻole Foundation defined a formal set of knowledge-organizing principles, or methodology, by referring to the Kumulipo. The Kumulipo is a *koʻihonua*, cosmogonic genealogical chant, that recites birth order. It is over two thousand lines long; has sixteen *wā*, spatial/temporal interval in which something is born; and is separated into seven *wā* for *pō*, darkness or night, and nine *wā* for *ao*, light or day. Humans enter the world in *wā ʻewalu*, section eight.

In *wā ʻumikūmākolu*, section thirteen, Kanahele found three words each on their own line (emphasizing significance): *papahulihonua, papahulilani,* and *papanuihānaumoku. Papa* is a foundation, stratum, flat surface, or layer. *Huli* is to search for, to research, to study. *Honua* is the physical entity known as the earth, *lani* is the physical entity known as the sky, *nui* is great or large, *hānau* is to give birth, *moku* is an island, district, or land division. So, *papahulihonua* is the knowledge foundation studying the earth and ocean, including its development, transformation, and evolution by natural causes. *Papahulilani* is the knowledge foundation studying the sky, the space above our heads to where the stars sit in the atmospheric expanse. *Papanuihānaumoku* is the knowledge foundation studying the birthing processes of all flora and fauna including humans.

Kanahele and colleagues have determined that these three words define three houses of knowledge that envelop the Hawaiʻi universe, and consider them the foundational layers or building blocks of all life and existence. They call this methodology Papakū Makawalu and describe it as "a way of learning a diminutive component while having some perspective of the full extent of the whole" (Kanahele et al. 2009, 32). *Papakū* is an established foundation of both tangible and intangible entities bound together and standing as one. *Makawalu* is an infinite movement, evolution, and transformation of the eight eyes. The eight eyes is a metaphor or symbol of chiefly *mana,* which is considered the most supreme form of *mana* for a human.

Papakū Makawalu is a "foundation of constant growth" (Kanahele et al. 2009, 30–34). It is a process of accurately tracking cultural progressions while learning about the environment as completely as possible. The three houses of knowledge established by *kūpuna* centuries ago are still relevant today. One might consider each *papa* similar to various academic disciplines such as geology, geomorphology, oceanology, climatology, astronomy, biology, botany, or zoology. However, each foundation includes a class of experts who are spiritually, physically, and intellectually attuned to both the area of focus *and* its relationship to the other two *papa*. The best way to understand this is with a simple working example.

Let me begin by using the term *papahulihonua* as the foundation piece. Keep in mind this is not meant to be an exhaustive list, just an example of how the process works and why it is specific to each place. Since

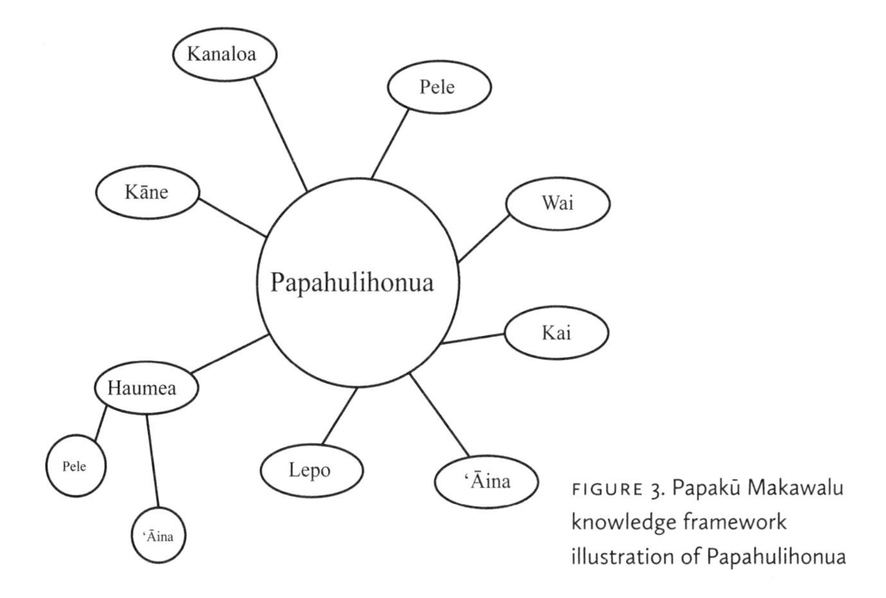

FIGURE 3. Papakū Makawalu knowledge framework illustration of Papahulihonua

papahulihonua is the study of the earth and ocean, the obvious "offspring" foundations or areas of study would be *ʻāina*, land; *kai*, sea water; *wai*, freshwater; and *lepo*, soil. However, we cannot forget the less obvious foundations, those processes of nature that *kūpuna* recognized and named as *akua*, such as Kāne, entity associated with life-giving processes such as freshwater processes; Kanaloa, entity associated with ocean elements and processes; Haumea, entity associated with life-forming and land-forming processes; and Pelehonuamea or Pele, entity associated with volcanic landforms and processes. Now here is where it gets interesting—*ʻāina* and Pele are also branches originating from Haumea, as shown in figure 3.

The difference between what the *ʻāina* or Pele represent in a Papakū Makawalu knowledge-organizing system depends on where it is placed in that structure, specifically, from where it originates. So, at the first level down from the main foundation of *papahulihonua* (the ellipses or flattened circles), both *ʻāina* and Pele are about studying the physical environment. So, *ʻāina* would be the study of *mokupuni*, island; *mauna*, mountain; or *awāwa*, valley, and Pele would be the study of volcanic landforms such as *lua*, crater, pit. However, when you are studying both *ʻāina* and Pele as part of the Haumea foundation (the circles) you are looking for those land and volcanic processes and not necessarily the features themselves. This is because they originate from Haumea and carry a component of Haumea, landform processes in its knowledge structure. So, *'āina* under Haumea would include the study of erosion, and Pele under Haumea would include the study of the movement of magma.

The brilliance and beauty of this system of knowledge is that any place or entity can be the starting *papakū*, foundation, from which you *makawalu*, branch out. For example, applying this methodology to a place like Hilo will produce different results from applying it to a place like Kona, although the structure will be similar. Each place will have specific land features, atmospheric observations, and living entities. This discourse can be applied to any feature or process as it underscores the connections all things have with earth, atmospheric, and life-generating systems.

Kanaka Hawai'i Sensory Perception

Kanaka Hawai'i knowledge about the world they experience is derived from the same sensory organs as every other person. Most Kanaka Hawai'i, like most people, have the same basic body structure, with two arms, two legs, and one head that has two eyes, two ears, a nose, and a mouth. Yet there seems to be something magical in the way they are able to look at the sky and predict an earthquake. Anthropologists E. S. Craighill Handy and Elizabeth Green Handy described it best when they wrote, "It is impossible to convey even a hint of the quality of mind and sensory perception that characterizes the human being whose perpetual rapport with nature from infancy has been unbroken. The sky, sea, and earth, and all in and on them are alive with meaning indelibly impressed upon every fiber of the unconscious as well as the conscious psyche" (1972, 23).

This heightened sensory perception many Kanaka Hawai'i developed is described in Meyer's work on Hawai'i epistemology and extended in Hawai'i geographer and language scholar Kapā Oliveira's work on Hawai'i senses of place. Although both acknowledge that Kanaka Hawai'i acquired spatial knowledge empirically through their body's sensory systems, Meyer emphasizes that a person's sensory inputs are culturally variable, and Oliveira points out that spatial knowledge is also genealogically embodied. In her interview with Rubellite Kawena Johnson, Meyer shares a Hawai'i understanding of six body-centric sensory inputs including the obvious five—vision, hearing, smelling, tasting, touching, and a conditioned intuitive awareness (2003b, 161–167). Reviewing the Hawai'i words for the five sensory systems reveals how Kanaka Hawai'i distinguish between physical capabilities of a sensory system and how the mind/body generates meaning from the sensory input.

For example, there are two Hawai'i words for seeing, *nānā* and *'ike*. *Nānā* is used when Kanaka Hawai'i want to look at something, to observe it, pay attention to it, or take care of it. *'Ike* is more than just seeing; it has to do with recognition, perception, and is used when Kanaka Hawai'i want to know or understand something. *Nānā* is the eye's physical capability, and *'ike* describes how the mind/body imbues what is seen with meaning (Pukui and Elbert 2003, 96, 260). "To hear" is the Hawai'i word *lohe*, and "to listen" is the Hawai'i word *ho'olohe*. The causative prefix *ho'o* changes the *lohe* from being something your ears are physically capable of doing to something your mind/body perceives. For Kanaka Hawai'i, listening requires interpretation beyond what the ear is capable of doing. For example, the sentence "*Ua hele 'oe i ke kula*" could be either the statement, "You went to school," or the question, "You went to school?" depending on intonation. Your ear can hear the difference between the two types of sentences, but your mind/body discerns the meaning of the intonation (Pukui and Elbert 2003, 209).

Kanaka Hawai'i differentiate between what the nose does, *hanu*, and how the mind/body interprets what the nose does, *honi*. *Hanu* has to do with breathing, respiration, and transpiration. *Honi* is to sniff or smell and is the word used to describe the way many Kanaka Hawai'i greet one another, by touching the side of their nose to the side of the other person's nose, during which both people inhale and share their breath or *hā* (Pukui and Elbert 2003, 57, 79). Kanaka Hawai'i understand eating is not the same as tasting. *'Ai* is to eat or take a bite and it is also food, specifically derived from non-meat sources. *Ho'ā'o* is the word Kanaka Hawai'i use for trying or tasting something (Pukui and Elbert 2003, 9, 27, 296). Kanaka Hawai'i also distinguish touching from feeling or groping. *Pā* and *ho'opā* describe physical contact, whereas *hāhā* is about gaining information from physical contact. A *kahuna hāhā* is a type of Hawai'i medical professional capable of making diagnoses such as sickness or pain simply by feeling or groping the body (Pukui and Elbert 2003, 46, 296).

Several of the terms used here to describe Kanaka Hawai'i sensory understanding have other meanings demonstrating the Hawai'i "logic by analogy" or "metaphoric mind." Furthermore, each of these five sensory systems has several other terms than those shared in the above text, indicating the degree of sensory refinement Kanaka Hawaiʻi achieved. According to Oliveira, Kanaka Hawaiʻi sensory refinement includes four genealogically embodied sensory systems, *naʻau*, ancestral lineage, *kulāiwi*, place lineage, *au ʻāpaʻapaʻa*, time/place/space, and *moʻo*, succession, all of which are considered something a child is born with or born into (2014).

Naʻau is commonly considered the sixth sense, intuition. This is similar to the conditioned intuitive awareness described by Johnson in Meyer's interview (Meyer 2003b, 161–167). Unlike the typical understanding of the sixth sense as something a person possesses by chance, Oliveira explains *naʻau* as originating from ancestral ties, like a spiritual umbilical cord serving as a generational link. *Kulāiwi* is generally understood as homeland and is described by Oliveira as a very deep and profound sense of place, both geographically and within the social hierarchy. However, *kulāiwi* is not just about relating to a particular place, but also linking to the knowledge from ancestors and elders who have developed cultural practices uniquely suited to each place and maintaining them for future generations.

Au ʻāpaʻapaʻa is presented as being firmly bound to a place for many generations—so much so, the place and the people become synonymous. Oliveira describes this as a sense of place/time, specifically how different places affect our perception of time, and vice versa, similar to the way different times in a person's lifetime affect their perception of place. Lastly, *moʻo* is expressed as succession and describes the sense of being the right person in the right place and at the right time to whom ancestral knowledge flows. Oliveira demonstrates how linguistic and metaphoric references inform Kanaka Hawaiʻi knowledge of a sensuous reality by illuminating the relationships between the physical, emotional, intellectual, and spiritual aspects of the Kanaka Hawaiʻi sensuous reality and revealing the power place ultimately has in influencing life and identity.

Our sensory systems provide us with information about the world around us. They mediate our experiences both through their structure and through the way we use each system. We shape our understanding of the world based on our experiences and our cultural conditioning. These perceptions, experiences, and cultural conditionings affect the way we orient our bodies, catalog our environment, and organize our enterprises in the world. Although this is but a very short introduction on a Kanaka Hawaiʻi cartographic philosophy, I believe there is enough here to begin to see how this perception of the world would generate distinctive cartographic practices. By personalizing their relationships with the natural environment and interacting with it daily, many Kanaka Hawai'i are able to access spatial knowledge instantly. From simply observing the environment, some Kanaka Hawai'i know when they see certain cloud formations or ocean tides it indicates an approaching storm, giving them plenty of time to make necessary preparations. This intimate interaction with nature shapes Kanaka Hawai'i cartographic practices. The next two chapters build on this philosophical underpinning, clarifying two important cartographic concepts, spatial/temporal orientation and classification.

CHAPTER TWO

Nā Kuana ʻIke Hawaiʻi

(Hawaiʻi Knowledge Perspectives)

At this point, we should be comfortable with the understanding that different cultures view the world through different cultural lenses. What may be sacred to one culture could be a commodity to another, and how we view the world dictates how we act in it. For Kanaka Hawaiʻi, there is no separation between the physical, social, intellectual, and spiritual aspects of Kanaka Hawaiʻi reality. That means we cannot completely understand the physical aspect of a place or entity if we exclude its social and spiritual counterparts. It also means that Kanaka Hawaiʻi believe acting in the physical realm without regard for the social and spiritual realms has consequences that we cannot understand or are not prepared to deal with.

Thus far we have learned a Kanaka Hawaiʻi cartographic philosophy based on *mana* as an ontological understanding that everything in this world has a life force, *ʻohana* as an epistemological framework of interrelatedness where everything is related to one another, *kinolau* and *kaona* as methodological processes of grouping characteristic likenesses and expressing a plurality of meaning, and Papakū Makawalu as a Kanaka Hawaiʻi methodology interconnecting foundations of evolutionary growth identified in the Kumulipo and shared by *kumu hula*, Hawaiʻi dance teacher, and Hawaiʻi scholar Pualani Kanakaʻole Kanahele of the Edith Kanakaʻole Foundation. We've also learned how knowledges are mediated through culturally conditioned and genealogically embodied sensory systems. This chapter builds on this philosophical foundation, presenting Kanaka Hawaiʻi cartographic knowledge embedded in spatial/temporal orientation. Let's begin with this integrated concept of space and time, which is similar to Norton-Smith's circularity as a world-ordering principle (2010, 10).

Kanaka Hawaiʻi cartographic practitioners accommodate for the integration of the abstractions of space and time. Instead they recognize space/

time as both an entity and a process that exists in this reality. The words *wā* and *manawa* are explained as both a space or interval between two objects or events and a period of time. Haumea is the entity Kanaka Hawaiʻi associate with fastening together space/time. Haumea represents the promulgation of life and ensures events occur in the right space at the right time. Have you ever noticed how planting two fruit trees in two separate locations on a sunny spring afternoon may result in growth disparities between the trees? The simple truth is that the right conditions are necessary for the fruit tree to be bountiful, such as soil type, lighting, and water accessibility. For Kanaka Hawaiʻi, the term "the right conditions" is the purview of Haumea.

Presenting spatial/temporal orientation in an organized and easily comprehensible fashion requires a discussion of scale. Kanaka Hawaiʻi had to create a word that describes scale as it relates to measurement because our relationship with the world did not involve mathematical fractions of numbers.[7] This is not to say that measurement did not exist or was not important to Kanaka Hawaiʻi. In fact, when exact measurements were necessary for specific purposes, like building a house, a special *ʻaha*, coconut braided sennit cord, was used to ensure the corner posts were properly placed for structural integrity. Kanaka Hawaiʻi also have a system of measurement, explained in the next chapter, on spatial/temporal classification.

For Kanaka Hawaiʻi, scale is relative to the conceptual level of understanding being referred to—the body, the island, or the planet. However, this concept of scale does not mean each is independent of the other. In fact, Kanaka Hawaiʻi cartographic philosophy dictates that they are connected or interrelated and that there are similarities and interactions between them. These next sections are meant to make it easier to understand the Kanaka Hawaiʻi cartographic perspective of orientation at various scales, and not to separate them from one another.

Kū i ke Kino (Pertaining to the Body)

Orientation at the "body as universe" scale is not just about human bodies for Kanaka Hawaiʻi. It includes all bodies, cloud bodies, rain bodies, fish bodies, bird bodies, even rock bodies. Each of these bodies is the center of its own universe and a means for understanding relative orientation. Using the human body as an example, our head is *luna*, up; feet are *lalo*, down; face is *alo*, front; back is *kua*; and sides are *ʻākau*, right, and *hema*, left. Many bodies are able to move, to negotiate through space/time. We designate relative direction according to the orientation of a single body, such as *mua*, forward, in front of, before; *hope,* backward, in back of, after; *waena,* between, in the middle of, among, in the center of; *loko,* inside of, within; and *waho,* outside of. We can also designate relative movements or positions based on the proximity or location of one body in relation to another body, such as *aʻe*, upward, on top of, above, on; *iho*, downward, on bottom of, below, under; *aku*, away from the body; *mai*, toward the body; *ʻō aku*, on the far side of, beyond the body; and *ʻaneʻi mai*, on the near side of the body (Pukui and Elbert 2003).

The terms *mua* and *hope* were introduced in *Aia i Hea Au?* (Where Am I?) with the concepts *ka wā ma mua*, the space/time in front of the body, and *ka wā ma hope*, the space/time in back of the body. These concepts also show that space/time is also a "body" to which relative direction can be applied. Oliveira expresses the relationship of these directional terms in regard to space/time with a series of images and examples (2014, 81–84). The text below extracts and expands some examples specifically focusing on the proximity of space/time to a body.

kēlā pule aku (that far week after)
kēia pule aʻe (that next week)
kēia pule iho (this coming week)
kēia pule (this week)
kēia pule iho nei (this past week)
kēlā pule aku nei (this last week)

As shown in Oliveira's diagram (2014, 81), *iho* occupies the closest proximity or immediate vicinity of the space/time body known as "the present." It can refer to either the immediate future or the immediate past with the addition of the term *nei. Nei* is a chameleon of sorts, as it has many meanings in relation to space/time depending on what it accompanies. It is used as a present tense marker because Hawaiʻi language does not generally change the form of a verb, it adds markers to indicate past, present, or future tense. For example, *e hele ana,* will go; *ke hele nei,* is going; *ua hele,* went. If *nei* follows a noun or pronoun, it adds a term of endearment such as *Hawaiʻi nei*, this beloved Hawaiʻi. Lastly, after directional terms, as shown in the example above, *nei* indicates past time.

Aʻe is further from the immediate vicinity of "the present" and refers to the immediately succeeding space/time unit. Although it appears to refer only to the future in the examples shown above, Oliveira explains that it can also refer to an older sibling, *ʻO ia koʻu mua aʻe*, s/he is before above me (2014, 82). However, I think the use of the term *mua* in this example means it has as much to do with the relative location of one body in relation to another body within a family unit as it does with the relative direction between the two bodies, or siblings being born in succession of one another as opposed to all the other older siblings. Thus, an older sibling could say of a younger sibling born in direct succession, *ʻO ia koʻu hope iho*, s/he is after below me.

Aku is the furthest from the vicinity of the space/time body known as the present, as it represents that direction away from the body. It is like *iho* in that it can refer to either the past or the future depending on the use of *nei*. It is most often used with verbs to indicate direction and is similar to its counterpart, *mai*, in that respect, often finely tuning the verb. For example, the verb *aʻo* is loosely explained as both learning and teaching. Although most teachers will agree that learning does indeed occur while they are in the process of teaching, the relative directional terms *mai* and *aku* distinguish *aʻo mai*, learning, from *aʻo aku*, teaching.

All of this discussion does not mean Kanaka Hawaiʻi do not have a system for ordering space/time. According to Hawaiʻi political scientist George Kanahele, Kanaka Hawaiʻi do understand the concept of time. It is just not as important as the activities, experiences, and events that occur in a person's life, which reside as timeless manifestations or memories within our mind/bodies (1986b, 168). Kanaka Hawaiʻi understand space/time as intertwined entities and order them according to the diurnal and seasonal cycles of the celestial bodies (covered in the section of this chapter "Pertaining to the Planet").

Kū i ka Mokupuni (Pertaining to the Island)

Orientation at the "island as universe" scale differs from orientation at the "body as universe" scale not only because of size, but also because of the nature of the two bodies. The Islands of Hawaiʻi appear to be stationary in the middle of the largest ocean on the planet. However, we now know that they are not stationary and that they do move very slowly toward the northwest. So slowly, in fact, that had the islands been populated when humans began walking the earth about two hundred thousand years ago, we still would not have recognized the islands were moving, as the youngest island, Hawaiʻi, is approximately five hundred thousand years old. So, while an island has a few relative location and relative direction terms, the apparent stationary nature of the island allows for the added layer of understanding absolute direction.

Beginning with relative location on an island, we have *mauka,* mountainward, or toward the mountain, and *makai,* oceanward, or toward the ocean. Relative location and relative direction away from an island includes both the *hālāwai,* horizon, and the *hoʻokuʻi,* zenith. Determining absolute location and direction at the "island as universe" scale involves interaction with *kalanipaʻa,* the celestial sphere, and the apparent movement of celestial bodies, that is, sun, moon, planets, stars. Today we understand that celestial bodies only appear to rise and set because the observer's point of view seems to be fixed. In actuality, the observer's point of view is not fixed. The observer is really on a rotating wobbling planet revolving around the sun in a solar system situated in one of many planetary tentacles spiraling out from a central origin in space. So, when observing the night sky from a seemingly fixed location, *kalanipaʻa* appears to move from east to west because the earth rotates from west to east.

While I do not know for certain whether or not Kanaka Hawaiʻi understood that it was the earth that was actually rotating and not *kalanipaʻa,* I do know that they inherited astronomical observations from the earliest Polynesians who entered the Pacific Ocean in 1500 BCE. The only absolutes in *kalanipaʻa* are that all celestial bodies consistently appear to arrive or rise from *kūkulu hikina,* the east. The term *kūkulu* means a point or end of the earth and *hikina* is a form of the term *hiki,* to arrive. All celestial bodies reenter the realm below the horizon at *kūkulu komohana,* west. *Komohana* is a form of the term *komo,* to enter. Malo lists another more poetic set of terms for east and west as *ka lā hiki* and *ka lā kau,* respectively. Since *lā* is a term for the physical object known as the sun and *kau* is to place, set down or to rest, the east is where the sun arrives and the west is where the sun sets down. Malo clarifies that these terms were specifically applied to borders or coasts of an island and not to *kalanipaʻa* (1971, 9–10).

According to Kepelino Keauokalani, there are two terms for north and south. The older terms are *koʻolau,* windward, for north, and *kona,* leeward, for south (1932, 80). While these older terms appear to be based on wind patterns, a case could be made that these terms have some connection to a Sāmoan perception of their island reality, as the terms are Hawaiʻi language versions of the island groups Tokelau and Tonga, which are northeast and southwest of Sāmoa, respectively. The newer terms for north and south, *kūkulu ʻākau* and *kūkulu hema*, respectively, may have more to do with the position or movement of the sun. *ʻĀkau,* where *ʻā* is in the nature of the noun that follows and *kau* is summer season, describes the direction where the sun resides during summer and the days grow longer. *Hema* is considered the opposite of *ʻākau* and describes the direction where the sun resides during winter and the days grow shorter.

The terms *kūkulu ʻākau* and *kūkulu hema* demonstrate Kanaka Hawaiʻi knowledge of the horizontal movement of the sun. When Kamehameha Investment Corp (KIC), a for-profit arm of the Kamehameha Schools Bishop Estate (KSBE), began restoring three *heiau* on the Keauhou, Kona, shoreline in 2007, they did not know one of the *heiau*, Hāpaialiʻi, was a solar calendar. The word *heiau* popularly designates stone temples or places of worship and is also understood by some Kanaka Hawaiʻi as the space/time where it is possible to ensnare the movement of energy (Kanahele et al. 2009, 14). The restoration of Hāpaiʻaliʻi was completed in December 2007, and a few weeks later, on the winter solstice, the sunset viewed from the center stone aligned with the southwest corner of the *heiau*. On the vernal and autumnal equinox, the sunset lines up in the center of the *heiau*, and the summer solstice lines up with the northwest corner.

The path the sun takes across *kalanipaʻa* during the vernal and autumnal equinox is called *ke alanui i ka piko o Wākea,* the great path of the navel of Wākea. Its counterpart on earth is *pōʻaiwaenahonua,* the equator. The sun's farthest northward progression during the summer solstice is *ke alanui polohiwa a Kāne,* the large dark glistening path of Kāne. On the earth this correlates to *pōʻaiʻoluʻākau*, the Tropic of Cancer. The sun's farthest southward progression during the winter solstice is *ke alanui polohiwa a Kanaloa,* the large dark glistening path of Kanaloa. On the earth it correlates to *pōʻaiʻoluhema*, the Tropic of Capricorn. Lastly, the eastern sky was considered *ke ala ula a Kāne,* the flaming path of Kāne, and the western sky was *ke alanui maʻawe a Kanaloa,* the great wispy path of Kanaloa (Poepoe 1906).

Tracking the sun was an important practice for Kanaka Hawaiʻi, as demonstrated with placement and *poho* (depression, hollow, cupule,

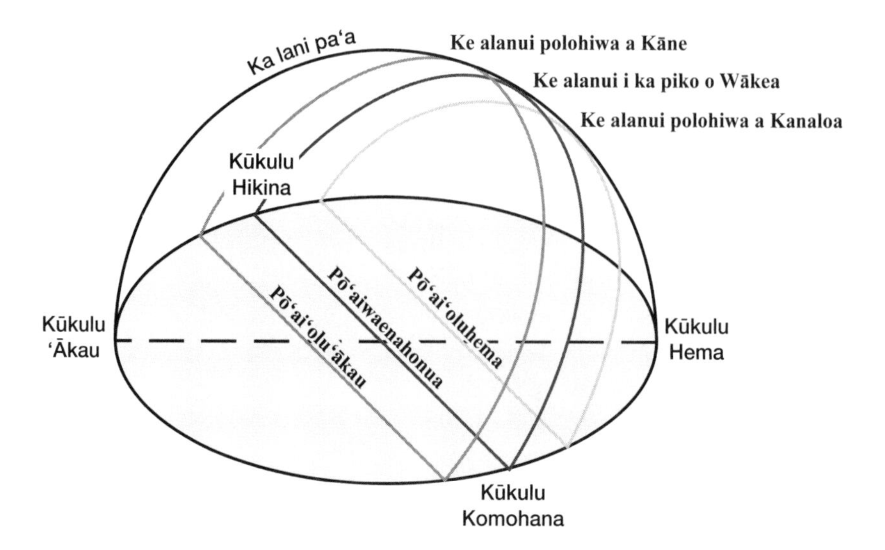

FIGURE 4. Kānaka Hawaiʻi Celestial and Terrestrial Absolute Location Designations

carvings) on Pōkāneloa, a rock on the island of Kahoʻolawe. Kahoʻolawe is situated to the west of Maui. Once occupied by the US Navy as a live-fire training area, it was returned to the State of Hawaiʻi in 1994 after a decade-long struggle by the people of Hawaiʻi, some of whom formed the Protect Kahoʻolawe ʻOhana (PKO). Anticipating the island's return, the United States created the Kahoʻolawe Island Conveyance Commission (KICC) in 1991 to make recommendations to the US Congress regarding the future uses of the island. The commission hired several consultants to research various aspects of Kahoʻolawe in order to obtain the most comprehensive background about the island. Pōkāneloa is described in Johnson's report ("Kahoʻolawe's Potential Astro-Archaeological Resources") as possibly having an astronomic function that merits further investigation (Johnson 1993). It took twenty-five years for a formal investigation to take place.

In 1993 the State of Hawaiʻi passed legislation designating Kahoʻolawe and the waters within two miles of it as a reserve and creating the Kahoʻolawe Island Reserve Commission (KIRC). The navy retained control of who could access the island until November 2003, as they commenced with the cleanup of unexploded ordinance. Since funding for the cleanup could not ensure complete removal of unexploded ordinance, KIRC worked with the Hawaiʻi community to designate areas of cultural, ecological, and historical significance that needed to be safe for future public interaction. Regardless of the existing safety threats and abuse shown by others, Kanaka Hawaiʻi embraced the island with the loving care given to any family member who had suffered such trauma. This healing process has been tended to for over a decade and comprises several restoration projects that integrate Kanaka Hawaiʻi cultural principles, protocols, and practices.

In March 2008, twenty-five years after Johnson had designated Pōkāneloa as a site needing further investigation, several Kanaka Hawaiʻi cultural practitioners were given access to the island to conduct a study of the sun at several locations on the island during the vernal equinox. The team who studied the sun at Pōkāneloa not only observed how the *poho* on the rock correlated with the sun's movement, making it a Hawaiʻi sundial, but also learned that one of the *poho* marked the setting angle of the descending sun. Most people know that, because of the tilt of the earth, the sun appears to rise from different locations along the horizon depending on the season. But not many people know that all celestial bodies, including the sun, do not ascend and descend perpendicular to the horizon unless you are at or near the equator. Celestial bodies ascend and descend at an angle.

This occurs because the planet is tilted on its north–south axis. The angle of tilt is equal to the latitude of the observer. Since Hawaiʻi is at about 20 degrees north latitude, all celestial bodies, including the sun, moon, and stars, appear to rise at 20 degrees to the right or south of a line perpendicular to the horizon and appear to set at 20 degrees to the left of a line perpendicular to the horizon. As the sun travels overhead we generally do not notice that it is traveling on an inclined path until it nears the horizon. At a certain point in the celestial sphere, the sun appears to travel in a perpendicular line.

The team of Hawaiʻi cultural practitioners watched as the shadow cast from a stick placed due west on Pōkāneloa during sunset on the vernal equinox aligned with *poho* no. 1 through *poho* no. 4 along the eastern edge. The final shadow aligned with *poho* no. 5, which is closer to the center of Pōkāneloa, indicating Kanaka Hawaiʻi captured the angle of sunset, shown in figure 5 (Kanahele et al. 2009, 5). Several other *poho* have yet to be correlated with the movement of the sun, indicating that there is still so much more to learn from Pōkāneloa. Thankfully, KIRC sought and secured a

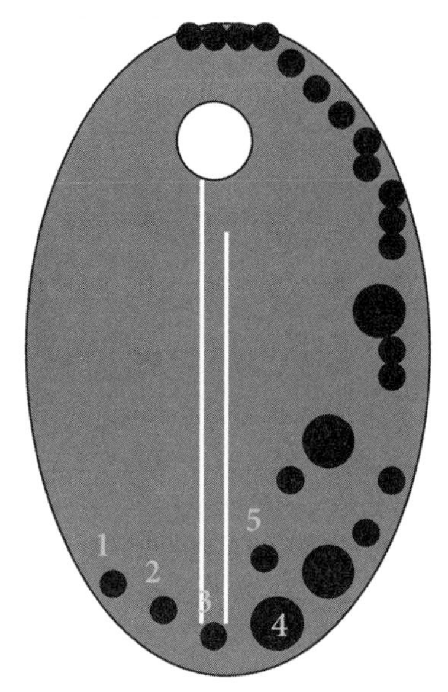

FIGURE 5. Pōkāneloa schematic diagram with locations of cupule. (Kanahele et al. 2009, 5)

one-time allocation of funds in fiscal year 2014 to purchase erosion control materials and plants to stabilize Pōkāneloa for future study and cultural use (Kahoʻolawe Island Reserve Commission and State of Hawaiʻi 2014, 20)

Kū i ka Honua (Pertaining to the Planet)

Orientation at the "earth as universe" scale is different from the previous two scales because Kanaka Hawaiʻi were physically bound to the surface of the earth and could observe only the effects of the planet's movement. For example, its rotation provides diurnal cycles, the moon revolving around it provides lunar cycles, its tilt provides seasonal cycles, and its revolution around the sun provides annual cycles. In other words, the cumulative effect of the "earth as universe" orientation leads to various cycles of life and is better known as the concept of "passing time." Furthermore, some Hawaiʻi language terms that describe the passing of time are based on observations of the movement of celestial entities, as you will see in the first example of the diurnal or daily cycle.

Diurnal cycles

Some Hawaiʻi language terms for the passing of daily cycles of time depict various characteristics of the sun, sea, and/or sky as the sun, the largest entity in *kalanipaʻa*, passes through different spaces within *kalanipaʻa*. Nuʻuhiwa discovered three different versions of when the day begins: sunset-to-sunset, midnight-to-midnight, and sunrise-to-sunrise. She asserts they are all correct depending on the position of the moon and the purpose of the activity associated with the particular day (2011). Please note, the Hawaiʻi language terms shared in the next few paragraphs are not meant to be a complete compendium of terms describing the "passing of time" on a daily cycle. Another Hawaiʻi scholar may describe this cycle using different terms or provide different explanations for the terms presented. These terms were selected because my metaphoric understanding of their explanations shows a connection between various characteristics of the sun, sea, and/or sky as the sun moves across *kalanipaʻa*.

Using the sunrise-to-sunrise version as an example, predawn is *paoa*, a rare variation of *pawa*, when the light from the sun breaches the horizon and lightens the darkness of night. Dawn or daybreak is *kaiao* or *wanaʻao*, where *wana*, to appear, and *ao*, light, describes the sunlight that first streaks across the sea. Morning is *kakahiaka*, where *kakahi*, solitary, unique, outstanding, and *aka*, shade or shadow, describes the sun that is high enough to cast shadows. Morning is further divided into early morning, *kahikole*, and midmorning, *kahikū*: early morning, *kahikole*, where *kahi*, place, and *kole*, red, swollen, enflamed, describes the space/time where the sun emerges out of the ocean and appears to be a swollen red orb; midmorning, *kahikū*, where *kū*, to rise, stand erect, describes the space/time when the sun stands above the horizon without the support of the sea.

Late morning to early afternoon is *awakea*, where *ākea*, broad, vast, remote, to be extended in breath, describes the space/time where the sun nears its apex and seems to slow down and metaphorically appears to be taking an extended breath. This is similar to the trajectory of a ball lobbed into the air—as it reaches its apex it moves slower than it does on the way up or down. Noon is *kau ka lā i ka lolo*, where *lolo*, brain, describes the space/time where the sun rests on the brain. Late afternoon is *ʻauinalā*, where *ʻauina*, descend, describes a space/time where the sun appears to decline more quickly. Evening is *ahiahi*, where *ahi*, fire, red, describes a space/time where the sun swells as it nears the horizon and the sky turns red. Dusk is *pōlehulehu*, where *pō*, darkness, night, obscured, *lehu*, ash, and *lehulehu*, multitude, describes a space/time where the fiery red sky is extinguished, leaving the afterglow on the horizon similar to ashes of a fire so the multitude of night can begin.

Night is *pō*, when sunlight is obscured. Midnight is *aumoe,* where *au* is period of time, current, flow, movement, and *moe*, to sleep, describes a space/time where the world sleeps. After midnight is divided into *ʻauinapō*, when night is in decline, and *pilipuka*, where *pili*, near in proximity, and *puka*, a hole, to pass through, appear, describes the space/time where the sun gets closer to emerging from the eastern horizon and light begins to brighten the darkness of night in the east. This term indicates Kanaka Hawaiʻi recognize that light travels faster than the sun because the night sky changes before the sun breaches the horizon. These descriptive terms for the passing of time are not only a testament of Kanaka Hawaiʻi observational acuity, they also affirm the space/time concept Kanaka Hawaiʻi embrace as time is described according to the location of the sun in the *kalanipaʻa.*

Lunar cycles

The next largest entity in *kalanipaʻa*, the moon, marks the passing of time at the lunar scale. In her presentation on Kaulana Mahina, the Hawaiʻi lunar calendar, Nuʻuhiwa (previously Tsuha) identifies thirty *pō mahina*, moon phases, organized in three *anahulu*, a ten-day period similar to a week: *hoʻonui*, waxing moons; *piha poepoe*, round and full moons; *and hōʻemi*, waning moons (see table 1). Each moon was named according to its effect on plant and fish life cycles, two ecological areas necessary to sustain Kanaka Hawaiʻi life (Tsuha 2008).[8]

TABLE 1: Nā Pō Mahina compiled by Nuʻuhiwa

Hoʻonui	Piha Poepoe	Hōʻemi
Hilo	Huna	ʻOlekūkahi
Hoaka	Mōhalu	ʻOlekūlua
Kūkahi	Hua	ʻOlepau
Kūlua	Akua	Kāloakūkahi
Kūkolu	Hoku	Kāloakūlua
Kūpau	Māhealani	Kāloapau
ʻOlekūkahi	Kulu	Kāne
ʻOlekūlua	Lāʻaukūkahi	Lono
ʻOlekūkolu	Lāʻaukūlua	Mauli
ʻOlepau	Lāʻaupau	Muku

There are twelve or thirteen *malama*, months, in the Kaulana Mahina, and each of the four main islands has its own naming system, which was standardized in the 1970s by Mary Kawena Pukui with the help of several Kanaka Hawaiʻi scholars (see table 2). Nuʻuhiwa says Kanaka Hawaiʻi accounted for extra days by adding *malama pili*, related or coinciding month, every three to four years to recalibrate the lunar calendar. Nuʻuhiwa's research is still ongoing, so I honor her request not to describe this process any further here, to allow her the opportunity to present this information for her own scholarly pursuits.

TABLE 2. Nā Malama (the months) standardized in the 1970s by Mary Kawena Pukui with the help of several Hawaiʻi scholars

Nā Malama	**Nā Malama Kalenekalio**
Welehu	October–November
Makaliʻi	November–December
Kāʻelo	December–January
Kaulua	January–February
Nana	February–March
Welo	March–April
Ikiiki	April–May
Kaʻaona	May–June
Hinaiaʻeleʻele	June–July
Hilinaehu	July–August
Hilinamā	August–September
Ikuā	September–October

Seasonal cycles

The passing of seasons occurs because the earth is tilted and leans either its northern or southern half toward the sun as it revolves around it. Hawaiʻi experiences two major seasons based on atmospheric conditions, *hoʻoilo* and *kauwela*. *Hoʻoilo* is the wet season, where *hoʻo*, a causative prefix that causes the action in the verb that follows, and *ilo*, to germinate or sprout, describe a space/time where seeds begin their transformation. It is a space/time where rough seas, high winds, and heavy rains besiege the islands, making deep-sea fishing and planting less productive. This season begins in Welehu, and its activities are considered the responsibility of Lono, the entity associated with agricultural fertility.

Kauwela is the hot, dry season, where *wela*, hot, describes a space/time where heat permeates the atmosphere. It is a time when the ocean is vibrant and food is plentiful. This season begins in Ikiiki (see table 2), and its activities are considered the responsibility of Kū, the entity commonly associated with war, governance, taking action, and making critical decisions. According to the Papakū Makawalu team from the Edith Kanakaʻole Foundation, Kū represents the condition or quality of survival. It is the Kū essence that causes a fruit tree to grow in less than ideal conditions. Kū is that force in nature that provides each entity with the tenacity, determination, and persistence to endure, grow, and thrive. Where Lono ensures every entity has the external support necessary to procreate, Kū ensures every entity has an innate nature to persevere no matter what. Kū is that intangible essence that makes a seed want to germinate, a plant want to adapt to new environments, and a bird want to evolve into a new species (P. Kanahele 2014).

Annual cycle

Lastly, the apparent movement of the thousands of stars in *kalanipaʻa* marks the passing of time at an annual scale. The year starts in Welehu, in the season of *hoʻoilo*, with *makahiki*, where *maka*, eye, beginning, point of a spear, describes the space/time where the constellation Makaliʻi, the Seven Sisters or the Pleiades, appears to arrive on the eastern horizon during sunset (Tsuha 2008). *Makahiki* is a four-month celebration of life when work and war cease. Some astronomical observations were recorded in *moʻolelo*, historical accounts. Two Hawaiʻi scholars, Rubellite Kawena Kinney Johnson and Kekuewa Kikiloi, unfold the metaphoric logic embedded in the Kumulipo.

Johnson, a scholar of Hawaiʻi language and literature, uses her fluency to research Hawaiʻi astronomy and geography. According to her, the Kumulipo has clues of how Kanaka Hawaiʻi and their predecessors in Polynesia structured the space/time of the "earth as universe" and explains that the first seven *wā*, taken together as one section, separates living beings into categories, apparently classifying them into groups of related forms similar to scientific phyla, genera, and species but using a broader range of characteristic similarities such as structure, shape, and needs for existence. The next nine *wā* describe the natural/social environment of humankind and mimic the poetic structure of the first seven *wā* (2000, 29).

Johnson identifies key movement terms in the prologue of the *wā 'ekahi,* first era, shown below, that specifically focus on *au,* space/time, including *lā,* the physical entity known as the sun, *honua,* the physical entity known as the earth, *lani,* the physical entity known as the sky, and *malama,* the physical entity known as the moon (2000, 37).

"Kumulipo Wā 'Ekahi"

O ke au i kahuli wela ka honua
At the time when the earth became hot
O ke au i kahuli lole ka lani
At the time when the heavens turned about
O ke au i kuka'i aka ka la
At the time when the sun was darkened
E ho'omalamalama i ka malama
To cause the moon to shine
O ke au o Makali'i ka po
The time of the rise of the Pleiades
(Beckwith and Luomala 1981, 58, 187)

According to Johnson, this prologue states the theme and demonstrates Kanaka Hawai'i understood several cycles of time, including annual solar time, where the earth hitched along with or moved because of the sun; the ecliptic, the imaginary line that marks the annual path of the sun and the unfolding or the changing constellations of the night sky; lunar time, where the sun repeatedly lights up the moon; and the start of the agricultural calendar, associated with the rising of the Makali'i constellation on the eastern horizon (2000, 37).

Hawai'i anthropologist Kekuewa Kikiloi's comfort with the Hawai'i language and regular visits to the Papahānaumokuākea, the Northwest Hawaiian Island Sanctuary, expands our cosmic understanding of the Kumulipo, correlating it with the physical environment. He discovered that Mokumanamana, a low island in Papahānaumokuākea also known as Necker, lies very near *pō'ai'olu'ākau*. Kikiloi states, "This latitudinal pathway essentially separates the Hawaiian archipelago into two spatial halves consisting of (a) the main Hawaiian Islands up to Nihoa ('ao,' or light where the sun reaches a zenith point overhead), and (b) the Northwestern Hawaiian Islands from Necker to Kure ('po,' or darkness, where the sun does not reach a zenith overhead)" (2010, 17).

Kikiloi researched Hawai'i language versions of previously translated texts to uncover the metaphoric logic embedded within documented

Hawaiʻi oral histories, allowing him to provide a more robust insight into the importance early Kanaka Hawaiʻi placed on the movement of the sun. He asserts Kanaka Hawaiʻi tracked the sun's farthest northward progression to Mokumanamana and turned the entire island into a *heiau*, thereby connecting the temples previously discovered on Mokumanamana with the sociopolitical development of Kanaka Hawaiʻi chiefdoms. Kanaka Hawaiʻi chiefly *mana* is invigorated when *aliʻi* build *heiau* in places with cosmological significance and genealogical relevance.

This is the beauty of metaphoric texts. They are designed to transcend a single purpose. Often information lays dormant within them until someone investigates the connections in new ways and brings them to light. In this case, Kikiloi used the Kumulipo to identify *pō* as the land area north of *pōʻaiʻoluʻākau* for two reasons. First and foremost, because Kanaka Hawaiʻi did not regularly inhabit the islands north of Mokumanamana, but also because the essence of Kānehoalani, the entity associated with the life-giving processes of the sun, cannot shine directly down through the top of a person's head at noon, akin to the concept of enlightenment but with a distinctive Kanaka Hawaiʻi perspective.

This relationship between Kānehoalani and the top of a person's head brings back into focus the fact that Kanaka Hawaiʻi really did not separate spatial/temporal orientation in the manner I present them here. The connection between the entities at the scale of *kalanipaʻa* and the human body was considered very important, especially for *aliʻi*, Hawaiʻi political leaders, whose *mana* is affirmed by performing rituals in *heiau* that allow for the direct flow of the essence of Kānehoalani into their being, not just their body, not just their mind, not just their spirit, but also the combined entity of body/mind/spirit, or their being. It is like plugging in to the cosmic origin of our solar system. While *aliʻi* could recharge their chiefly *mana* through explicit rituals in specific locations, this did not eliminate others from also receiving a charge, as everyone has daily access to Kānehoalani and can charge the essence of their being with the noonday sun.

The next chapter builds on these Kanaka Hawaiʻi cartographic concepts of spatial/temporal orientation, direction, and perspective. It focuses on Kanaka Hawaiʻi spatial/temporal classifications using the same scales presented here. However, the same caution remains: these divisions of scale are arbitrarily assigned in order to discuss Kanaka Hawaiʻi cartographic concepts, and do not reflect any rigid or bounded construct. The structure is meant to inform, not to set in place as a fixed universality. This means some concepts are highlighted while others are not—no different from what a cartographer making a map for a particular purpose does. Some information is included in the map compilation while other information is not, in order to make the message or purpose of the map clear for the observer/participant.

CHAPTER THREE

Nā Māhele ʻIke Hawaiʻi

(Hawaiʻi Knowledge Classifications)

The previous chapter showed that a Kanaka Hawaiʻi cartographic understanding of the world at various scales directly reflects intimate observations and recordings of their environment. Some important ideas to remember is that orientation of the "body as universe" scale incorporated the concepts of relative location and direction, whereas orientation of the "island as universe" scale added the concepts of absolute location and direction. At the "earth as universe" scale, orientation led to the "passing of time" and demonstrated how and why Kanaka Hawaiʻi integrated the concepts of space and time.

Spatial/temporal classification is a natural extension to orienting the mind/body in space/time; however, this classification is not confined to the physical realm alone. It includes the social, political, and spiritually transcendent realms as well, and though concepts are presented according to scale for ease of understanding, they are not mutually exclusive, nor are the classifications meant to create bounded boxes that cannot be penetrated. Scale is used specifically for comprehension of Kanaka Hawaiʻi cartographic classifications. While it is easy to comprehend that a different lens will inevitably provide a different classification structure, it is also true that my sensual prescription makes my perception of classification specifically cartographic.

Kū i ke Kino (Pertaining to the Body)

Most people tend to classify the human body according to how each part functions. Our feet help us move about the earth; our arms help do things such as build, embrace, and create; our heart rules our emotions; and our head is usually considered the seat of logic and intellect, unless you are Kanaka Hawaiʻi and place the seat of intellect in your *naʻau*. But that is not the only distinction of how Kanaka Hawaiʻi cartographic practitioners classify the human body. First and foremost, most Kanaka Hawaiʻi do not view the human body as just a physical entity. They also recognize and classify the human body's ability to connect with the entities and processes in any space/time through three *piko*, end or extremity of an entity, such as the end of a rope, navel, or belly button: the *manawa*, *piko*, and *maʻi*. The *manawa*, fontanel, is a connection to the spiritual realm; the *piko* is a connection to genealogical ancestors; and the *maʻi*, genitals, is a connection to the next generation or our offspring. Understanding the human body as a node of connectivity allows various Kanaka Hawaiʻi cultural practitioners to further identify parts of the human body and its movements according to their specific training, such as *lomilomi*, Hawaiʻi massage; *lāʻau lapaʻau*, Hawaiʻi practice of addressing pains and illnesses with plants and prayer; and *hula*, Hawaiʻi dance.

Body-centric professional classifications

When I was studying *lomilomi*, I learned that aches on the right side of the body indicated the person was having issues with a male entity and on the left side indicated issues with a female entity. This could be either an external or internal entity, as many Kanaka Hawaiʻi believe we each have within us both maleness and femaleness, similar to Carl Jung's anima-animus archetype, wherein each individual is considered psychologically gender-neutral. This archetype is not the same thing as gender or sexual orientation. Nor does it mean there is no predetermined characteristic or trait assigned to being a man or a woman, such as assertiveness and tenderness. It is all about how an individual relates to these characteristics. A person's body will ache according to their perceived understanding. Ultimately, I learned that what happens within the body is a direct result of the person's interaction or lack thereof with what is perceived to be happening both internally and externally.

At Aunty Margaret Machado's Hawaiian Massage Academy, we had a workshop on *lāʻau lapaʻau*. According to the *lāʻau lapaʻau* practitioner who gave the workshop, picking medicinal herbs with the right hand meant something different from picking with the left hand. I later learned that this is not necessarily true with all *lāʻau lapaʻau* practitioners, as each practitioner has a distinct way of executing their training. For example, one practitioner may use a particular plant to reduce fever that another practitioner will say does not work. Yet each is considered a good practitioner because many Kanaka Hawaiʻi believe medicinal plants picked with the intentionality embedded into the practitioner via years of training will respond appropriately. In other words, if a plant is picked with the intention of reducing a fever, the plant will act accordingly. However, if there are leftovers of the plant and someone needs it to heal a broken leg, the plant will not work to its optimum potential because the plant allowed itself to be picked, literally giving its life up, for a particular purpose, reducing a fever.

Taupouri Tangarō, *kumu hula*, Hawaiʻi dance teacher, and Hawaiʻi mythology scholar, says the *ʻaihaʻa*,[9] a *hula* tradition maintained by Hālau o Kekuhi on Hawaiʻi Island, views the upper half of the body above the navel as the space above the earth and the lower half as the earth. Body movements above the navel represent entities such as trees, winds, flowers, and rain, and movements below the navel are connected with an animate earth. He also stresses that not only does each *hālau*, Hawaiʻi place of learning, have its own distinctive dance style, some *hālau* also assign their own distinctive meanings to similar movements. Furthermore, one style is not necessarily better than another, as each *hālau* maintains the traditions that reflect and embody the place from where it originates. Remaining true to tradition is paramount, but that does not mean *hula* cannot adapt and evolve, creating new dances that reflect ongoing interactions with the environment.

Oftentimes a person's function in society is dependent on societal classifications of our body, such as age, gender, status, and profession. Some Kanaka Hawaiʻi children are still selected at the early age of one to two years by a *kahuna*, Hawaiʻi master practitioner, because they demonstrate a particular predisposition or characteristic, and are trained to become masters of specific cultural practices. Other children are expected to help their *ʻohana*, family, and accept *kuleana*, privileged responsibility, depending on those societal classifications established during the *ʻaikapu*, era of restricted eating.

Body-centric gendered classifications

The *ʻaikapu* marks a period when the people in the Islands of Hawaiʻi created and regulated gendered divisions of space/time. It began during the reign of Wākea and his wife Papahānaumoku. Although they are not the "first" people to arrive in the Islands of Hawaiʻi, they are considered the principal progenitors of Hawaiʻi society because they made huge contributions to its sociopolitical structure. Their legacy has been encapsulated in various *oli,* chant; *kaʻao,* traditional story, tale of ancient times, legend or myth; *mele koʻihonua*, cosmogonic genealogy; and ceremonies.

One of their better-known *kaʻao* illustrates some of these sociopolitical contributions. In the early part of Wākea and Papa's union, they conceive a daughter and name her Hoʻohōkūkalani, which is generally explained as one who creates the stars, but in my metaphoric interpretation she is also the initiation of heightened intellect. As she matures into a beautiful young woman, Wākea begins to desire her intimately and seeks advice from his *kahuna,* Komoʻawa. Komoʻawa advises Wākea to declare some nights *kapu,* sacred or restricted, requiring men to conduct ceremonies away from women. On one of these *kapu* nights, Wākea convinces Hoʻohōkūkalani to break this *kapu* and lay with him intimately.[10] She becomes pregnant with their eldest child, a son they name Hāloanakalaukapalili. However, the child is born a *keiki ʻaluʻalu*, premature baby, and buried on the eastern side of the house from which a *kalo*, taro, plant later emerges. Later Wākea and Hoʻohōkūkalani conceive a second child together, a son they name Hāloa, in honor of his elder brother.

This *kaʻao* has at least three major implications for Hawaiʻi society. It introduced a class of *kāhuna*, plural form of *kahuna*; instituted gendered divisions of space/time; and established an ecologically integrated evolutionary genealogy of Kanaka Hawaiʻi. According to S. M. Kamakau, prior to the era of Wākea and Papa, Hawaiʻi was sparsely populated. Parents were the head of their own *ʻohana* and *kāhuna* were integral members of an *ʻohana* who became extremely skilled in a particular profession. After the era of Wākea and Papa, *kāhuna* became a class separate from the *ʻohana* and were respected as the authority on issues within their professional purview. "The separation began with the priesthood order of Lihauʻula, the first child of Kahikoluamea and older brother of Wakea [*sic*]. This order, the *papa kahuna pule*, was the first to be selected out (*wae*) and so the *kahuna* orders were kept separate throughout the entire race in the following generations" (1964, 3–4). The class of *aliʻi*, Hawaiʻi political leaders, followed some twenty-five generations later.

The creation of gendered divisions of space/time separated men and women for a majority of daily and ceremonial tasks. However, gender did not mean the same thing to Kanaka Hawaiʻi in the *ʻaikapu* era that it does today. At that time, Hawaiʻi was a polygendered society. Gender was not considered a binary biological phenomenon of *kāne,* male, and *wahine,* female; it was also a heightened holistic expression and included *māhūwahine,* a biological male who is woman-like, and *māhūkāne,* a biological female who is man-like (Snow 2014).

Māhū, a transgendered class of people, were integral parts of Hawaiʻi society, and although exempt from the rigid rules based on gender, they still had *kuleana. Māhūwahine* were allowed to remain with the women during male ceremonies. Knowledge of *māhū* is still evolving in modern research as we shed more than a century of denial, disdain, and derisive contempt from our colonial conditioning (see Tabag 2010). According to Hinaleimoana Wong-Kalu, a respected and beloved transgendered *kumu hula* and cultural scholar determined to carve a middle ground for future generations, *māhū* were well regarded as excellent caregivers for both the young and the elderly and often taught and maintained cultural practices such as craft-making, *lei*-making, chanting, singing, *hula*, storytelling, and genealogical maintenance (2001, 221). Some *māhū* continue some of these same practices today.

Body-centric genealogical classifications

The establishment of an ecologically integrated evolutionary genealogy with Hāloanakalaukapalili, the *kalo*-child, as the older sibling to Hāloa, the human child, embeds an ecological consciousness as part of the moral fiber in the felting of Hawaiʻi society. In an *ʻohana*, the older siblings are responsible for the caregiving and nourishing of younger siblings. Younger siblings reciprocate by listening to and supporting their older siblings. This familial practice reaches beyond what is considered the nuclear family and includes all the elements and processes of nature. Since plants, animals, and the elements and processes of nature existed before Kanaka Hawaiʻi, they are considered older siblings who provide the nourishment Kanaka Hawaiʻi need to survive. Likewise, as the younger sibling, Kanaka Hawaiʻi tune in to the needs of their elder sibling and provide the support it needs (Kameʻeleihiwa 1992).

This creates the perfectly balanced reciprocal relationship, wherein each party sacrifices a part of themselves for the other. The *kalo* surrenders itself to the Kanaka Hawaiʻi who cultivates, plants, waters, weeds, and feeds the plant. Each grows from the other's sacrifice. Many Kanaka Hawaiʻi consider sacrifice a good thing that can take many forms, from a simple *pule*, prayer, to the ultimate sacrifice of a life. As the elder sibling, it is the *kuleana* of the *kalo*-child to nourish its younger sibling, the human-child. From the perspective of Kanaka Hawai'i, every body, including those bodies that are not human bodies, has *kuleana*. Plants, animals, rain, wind, water, tides, and celestial bodies all have *kuleana*, and Kanaka Hawai'i classified each according to its characteristics and functions, just like they did their own bodies.

Kū i ka Mokupuni (Pertaining to the Island)

Many Kanaka Hawai'i understand that the elements and processes of nature existed before the human peopling of the planet and recognize the plants and animals in the ocean, on the land, in the atmosphere, and in the celestial sky are older peoples. While learning about each of these peoples through careful observation and intimate interactions, they also recognize how external influences affect growth. Medicine growing next to certain plants or in specific places enhances or diminishes its ability to heal. So it is beneficial to maintain knowledge at the "island as universe" scale in order to better maintain relationships with those natural elements and processes who are considered elder siblings.

Island-centric physiological landscape classifications

Most scholars relate the classification of the physical island environment as vegetative zones or ecological systems. Based on a Kanaka Hawai'i cartographic philosophical underpinning, I would like to add the concept of expressing a physiological classification of the islandscape. This is not an attempt to anthropomorphize the natural environment. This concept, anthropomorphize, comes from an anthropocentric tradition. Most Kanaka Hawai'i have more of a kin-centric tradition, and thus the term "physiological" is an accurate presentation of a people who maintained relationships with literally hundreds, if not thousands, of land, ocean, atmospheric, and celestial more-than-human bodies. Instead of simply listing the different physiological classifications, as Malo first published and Kamakau later expanded, or reprinting Oliveira's illustrations, I prefer to use my cartographic training and present physiological classification overlays on the oblique landscape images shown in figures 6 and 7.

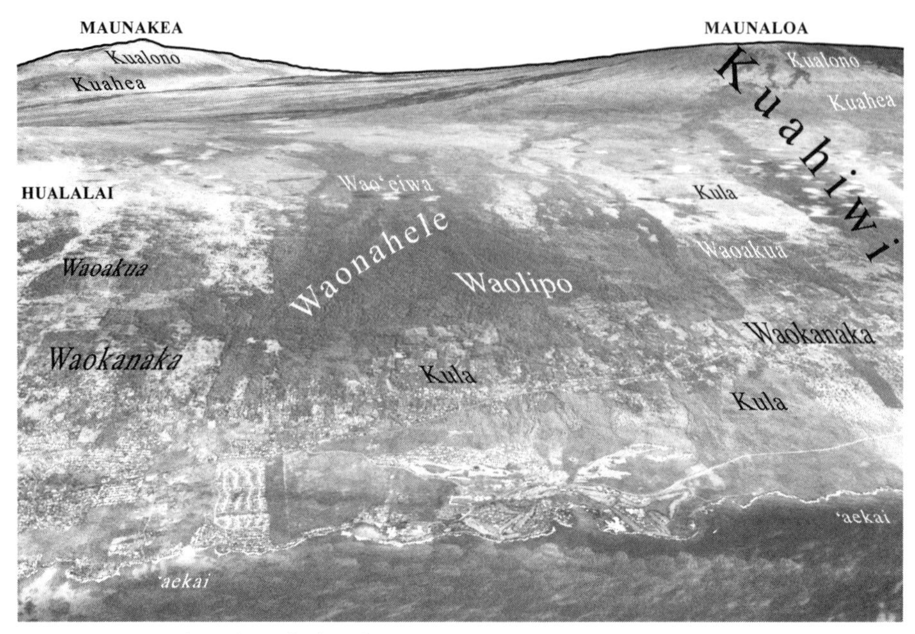

FIGURE 6. Kona Physiological Classifications

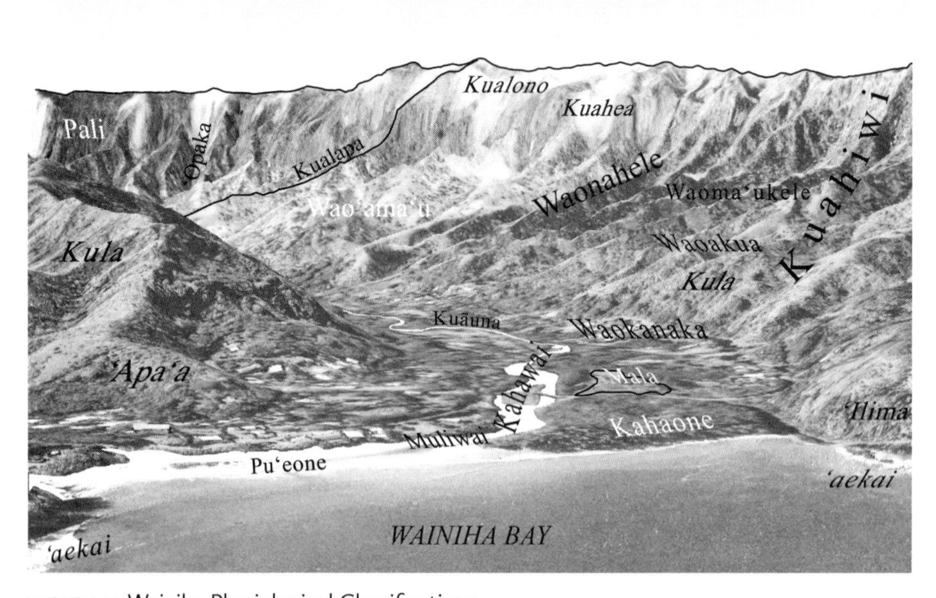

FIGURE 7. Wainiha Physiological Classifications

My predecessors list and illustrate the classifications as though they are in succession, but that does not seem to fit the specifics of a landscape. I chose the Kona or southwest side of Hawaiʻi Island and Wainiha or the north side of Kauaʻi for two reasons. They are at opposite ends of the Hawaiʻi archipelago and have distinctly different physiologies. Hawaiʻi Island is the youngest island and has not been as sculpted by the elements as the older island of Kauaʻi. Kona is drier and receives far less rain than Wainiha, yet there are ample inland areas providing an abundance of foods and other materials used in daily living. Wainiha is a well-developed valley with access to a wide range of growing conditions and an easily accessible bay at the shoreline.

Both locations have a few physiological similarities, especially in the mountain regions, but the interior wet forest, dry forest/grasslands, and coastal areas differ a bit. Both places have *kualono*, mountain region where the sounds of silence are deafeningly loud; *kuahea,* mountain area where mist forms and trees are stunted by altitude; and *kuahiwi,* mountain region extending from the *kualono* to where the land flattens. Most Hawaiʻi islandscapes have both *kualono* and *kuahiwi*. Metaphorically, I understand mountains as the *kua,* spine, of the earth, and descriptors such as *lono,* to hear, *hiwi,* skinny, bony, thin, angular, sharp ridge of a mountain, and *hea*, misty, clouded, obscured, add another level of understanding, allowing appropriate etiquettes of access to be followed. Some mountain areas are not easily accessible, such as those found in the well-sculpted Wainiha valley, including the *kualapa*, mountain ridge where *lapa* is a steep side of a ravine; *pali*, mountain precipice or cliff; and the *ʻōpaka*, mountain ravine. Mountain areas not depicted include the *kualipi,* sharp mountain ridge where *lipi* is a thin tapering edge like that of an axe; *kuanihi,* steep mountain ridge where *nihi* is a steep edge or border with difficult or precarious passage; and the *kuaola,* verdant mountain areas where all things grow and thrive, as *ola* is life.

The interior regions of both images include *wao,* wet forest area, and *kula*, dry grassland areas. The Kona image shows two types of wet forest areas, the *waoʻeiwa*, area covered with vegetation and small forest trees, and the *waolipo,* dense wet forest vegetation with tall forest trees, both of which are found in the *waonahele,* an inland wet forest area covered in vegetation, as *nahele* is wilderness. These interior physiological classifications are not evident in the *waonahele* of the Wainiha valley image because I concentrated on showing only a quarter of the length of the valley near the coastline. As the valley deepens into the interior of the island, these classifications become evident. What the Wainiha image does show is the *waoʻamaʻu,* forested region where ferns grow, and the *waomaʻukele*, forested rain belt where the ground is always wet and slippery.

The inland areas found on most Hawaiʻi islandscapes are the *waoakua,* forest region for the *akua,* divine entity associated with life-giving natural processes, and *waokanaka,* forest region for Kanaka Hawaiʻi. My metaphoric understanding of the *waoakua* is that it is the backbone of the forest. It is where the *wao* is a living entity that does most of its work and takes care of its various responsibilities, including capturing water in the atmosphere and feeding the tree peoples, fern peoples, and grass peoples so they continue cleaning the air and attracting water in the atmosphere. Any leftover water that is captured is stored in underground aquifers. Any disruption to this area severely hinders the ability of the *wao* to fulfill its responsibilities. This is why the majority of the Kanaka Hawaiʻi built or altered islandscape occurred in the *waokanaka*.

Each location also has *kula* regions, those open areas that are usually dry. In the Kona image there are two dry vegetation areas, one near the mountain and one near the coast. It is important to recognize that there is no fixed order to these physiological classifications. Each place brings a unique combination of conditions. While it is generally true that below the *kua* is the *wao* followed by the *kula* and ending with the *kahakai,* coastal region, it is not always the case. On the Kona side of Hawaiʻi Island there exists an open dry area above the *wao* that differs from the open dry area near the *kahakai.*

The Wainiha image has additional lowland regional physiological classifications of *ʻāpaʻa*, dry area with small trees, and *ʻilima*, dry shrublands. It also has three *wai,* freshwater, features, including a *kahawai*, freshwater stream, with a *kuāuna,* valley-like stream bank, and a *muliwai*, freshwater stream mouth. There is even a *mala*, cultivated dryland garden, large enough to give you the idea of what was once a valley floor filled with productive staples. The *kahakai* of the Wainiha image also includes a *puʻeone*, sand bar, and *kahaone,* sandy beach. Lastly, both images have an *ʻaekai*, sea water edge.

These images are not meant to give you the full extent of the physiological classifications in the Kanaka Hawaiʻi system. They are meant to demonstrate that different environmental conditions create different islandscapes that do not necessarily fit nicely into standardized classification schemes. Winds may create a dry and barren hillside on one side of a mountain that is lush and green on the other. Changes in the amount of rain may reduce or enlarge forest areas. Human development is one of the major activities altering the physiological classifications of the Hawai'i islandscape. However, the physiology of an islandscape is not just about the land; it is also about the ocean and sky.

Island-centric physiological oceanscape classifications

Kamakau relates nearly thirty ocean classifications that a Kanaka Hawai'i system distinguishes from the coast to the deep sea, almost as many wave classifications, and over a dozen tide classifications. Later, Harold Kent combined all three into one ocean category and listed over three hundred ocean terms in his compendium, *Treasury of Hawaiian Words*. The large number of terms is a testament not just to Kanaka Hawai'i intimacy with the oceanscapes surrounding the Islands of Hawai'i, but also to their intellectual capacity to discern the differences in what they experienced. Some islands have shorelines with long reefs where waves break far from shore and others have sheer cliffs that fall into the deep ocean close to the shoreline. Kamakau's ocean classifications suggest there are four distinct regions that are dependent on their distance from land, including the near shore, near reef, near land, and deep ocean.

However, keeping in mind that many Kanaka Hawai'i related to natural elements and processes as other-than-human peoples who have their own body of intelligence, my metaphoric mind understands three physiological classifications according to (1) distinguishing characteristics, (2) where other-than-human peoples reside, or (3) where Kanaka Hawai'i interact with other-than-human peoples. For example, a distinguishing characteristic of how the ocean meets the land is shown in the above figures as *'aekai*. But it can also be considered *lihi kai*, sea water edge, depending on the wave energy and tidal pull. Another example of a distinguishing characteristic classification near the shoreline is *pālaha*, spread out, broaden, flatten; or *pāhola*, to spread about, extend, diffuse; or *hohola*, to spread out, unfold, unfurl. These terms specify an area where the ocean water does not evaporate, sizzle, or crystallize as waves wash over the land, spreading the ocean water before it percolates down into the land. *Kai 'elemihi*, sea area associated with a type of crab, *kai kāhekaheka*, sea area associated with small salt-collecting pools, and *kai ki'oki'o*, sea area associated with water-collecting rocky basins, are examples of where other-than-human peoples reside.

Lastly, *kai heʻenalu*, sea area associated with surfing activity, *kai paeaea*, sea area associated with pole fishing activity, and *kai kāʻili*, sea area associated with hook and line fishing without a pole activity, are examples of ocean classifications according to Kanaka Hawaiʻi interactions.

Island-centric physiological skyscape classifications

Both Malo and Kamakau provide celestial physiological classifications. Malo lists two different sets of atmospheric physiological terms, *kahiki* and *lewa*. The term *kahiki* has many explanations. In the context of a skyscape entity's physiology it relates to "the arrival" of celestial bodies. *Lewa* is the lower regions of the skyscape where atmospheric bodies float, dangle, oscillate, or are otherwise suspended. Kamakau explains *kahiki* terms as different horizons where different lands are found and describes a different set of *lewa* terms for the strata of space, which is illustrated in Oliveira's text. It was easy to understand the atmospheric physiological classification of the *lewa* terms but not so with the *kahiki* terms until I cleared my mind of other people's explanations and sketched my own understanding.

It all came together when I read these three passages from Malo (translated by Nathaniel Emerson):

> As to the heavens, they are called the solid above, *ka paa iluna*, the parts attached to the earth are termed *ka paa ilalo*, the solid below; the space between the heavens and the earth is sometimes termed *ka lewa*, the space in which things hang or swing.
>
> According to another way of speaking of directions (*kukulu*), the circle of the horizon encompassing the earth at the borders of the ocean, where the sea meets the base of the heavens, *kumu lani*, this circle was termed *kukulu o ka honua*, the compass of the earth.
>
> The border of the sky where it meets the ocean horizon is termed *kukulu-o-ka-lani*, the walls of heaven. (1971, 9–10, italics original)

The spherical sketch shown in figure 8 is a creation combining my cartographic training with my metaphoric understanding of the above passages. I realized *kalanipaʻa* consisted of an outer and inner dome, which could possibly be associated with *ka paʻa i luna* and *ka paʻa i lalo*, respectively. The *kahiki* terms apply to the arc of the outer dome and the *lewa* terms apply to the inner dome. This allowed my metaphoric mind to understand both the *kahiki* and *lewa* terms as characteristics of the celestial entities in different portions of the celestial sphere.

Kahikimoe, where *moe* (sleep) describes that part of the celestial sphere where celestial entities appear to be lying on the surface of the earth, slowly rising from sleep. Here the tilt of the earth makes it appear as though the celestial bodies are slowly rising and are not yet able to stand. *Kahikikū*, where kū (to stand upright) describes that part of the celestial sphere where celestial bodies stand without the support of the surface of the earth and no longer appear to be traveling on a slanted path. *Kahikikapapanuʻu*, where *nuʻu* (to rise up) describes that part of the celestial sphere where celestial bodies ascend. *Kahikikapapalani*, where *lani* (the physical entity known as the sky) describes that part of the celestial sphere where celestial bodies enter the upper echelons and then appear to slow as they move across the sky. *Kahikikapuihōlaniikekuina*, where Hōlani is a mythical place and *kuina* (a meeting place) describes that part of the celestial sphere nearest the apex or zenith.

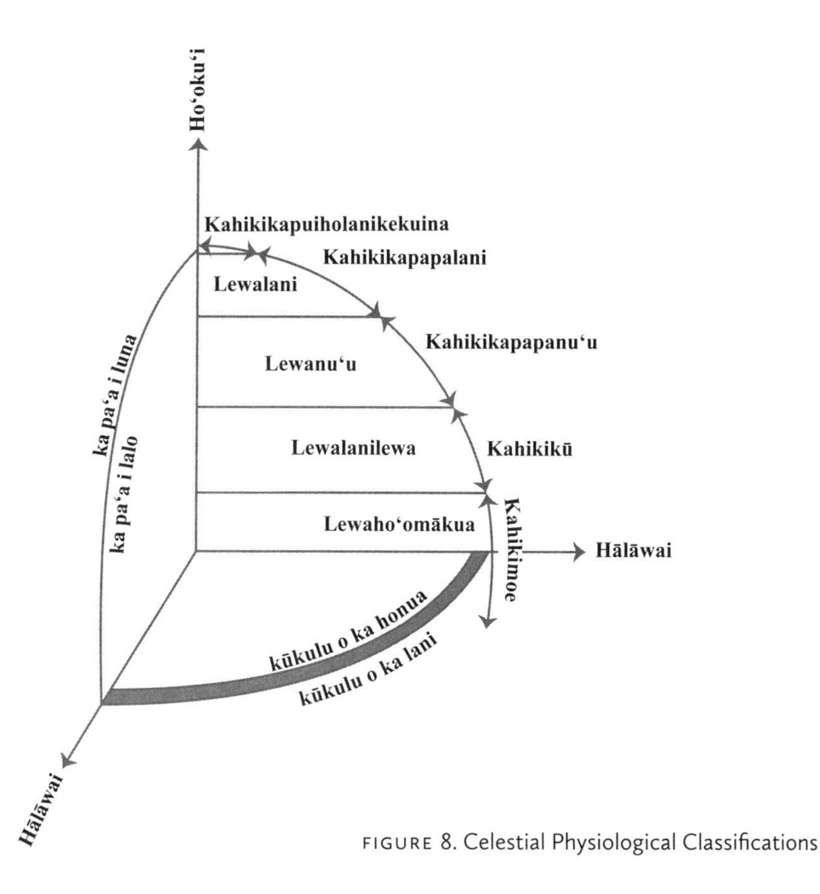

FIGURE 8. Celestial Physiological Classifications

Where *kahiki* terms describe the characteristics of celestial bodies along the arc of the celestial dome in *ka paʻa i luna*, the *lewa* terms describe the characteristics of bodies within the bounds of the celestial dome in *ka paʻa i lalo*. My metaphoric understanding of *lewahoʻomākua*, where *mākua* (full grown, established, to be large, grow, strengthen, sustain) is the atmospheric layer where many bodies continue to multiply, thrive, and flourish. The next atmospheric level up is the *lewalanilewa*, where low-hanging clouds move across the sky and occasionally kiss the surface of the highest mountains. The *lewanuʻu* is where birds fly, and *lewalani* is the highest layer of the atmosphere, where a majority of atmospheric bodies, both entities and processes, reside, such as the various winds, rains, and cloud formations.

It is important to note that the sketch in figure 8 and explanation of the celestial dome are not completely consistent with nor corroborated by other modern scholars and may not be a correct representation of Kanaka Hawaiʻi understanding. I consider this alternative expression one of several interpretations that, like the facets of a cut gem, will help reveal the beauty within this cartographically framed rationalization and lead to a more refined Kanaka Hawaiʻi understanding of atmospheric and celestial classifications.

Island-centric physiological starscape classifications

Hawaiʻi language newspapers have published several articles on *hōkū*, star, classifications. Through the efforts of thousands of volunteers on the ʻIke Kūʻokoʻa project, these newspapers are not only accessible on the Internet but are also searchable via the Office of Hawaiian Affairs Papakilo database. From these online searchable sources, the most comprehensive star classification list comes from the genealogy or ancient history of Kanalu. According to Kanalu there are eight classes of stars, including *hōkūaliʻi*, royal stars; *hōkūmakaʻāinana*, plebian stars; *hōkūhōʻike*, prophetic stars; *hōkūkahuna*, stars for Hawaiʻi priests; *hōkūʻāina*, land stars; *hōkū no ke akua*, stars relating to the god; *hōkū no ka malama*, stars for every month of the year; and *hōkūkilo*, stars usually observed by Hawaiʻi astronomers. Unfortunately, Kanalu does not list the names of individual stars or star constellations for each category (Condis 1909a). Kamohoula contributed over thirty-six star names in three categories, but did not indicate how they were organized, only whether they were in the first, second, or third classification (Makemson 1939; Condis 1909b, 1909c, 1909d, 1909e, 1909f, 1909g, 1909h).

In his own publication, Kepelino Keauokalani indicates there are two main groups of stars: *hōkūpaʻa,* fixed stars, and *hōkūlewa,* moving stars. He further identifies three types of *hōkūpaʻa* and two types of *hōkūlewa.* The *hōkūpaʻa* group includes *kiopaʻa,* the North Star; *lalani,* the Milky Way; and *hōkū ʻaeʻa,* wandering stars, also known as planets. The *hōkūlewa* group includes *hōkū kiaʻi,* guiding or guardian stars, used by navigators, and the *hōkūlewa ʻano ʻole,* insignificant moving stars (1932, 78–83). At first glance it seems counterintuitive to list both *lalani* and *hōkū ʻaeʻa* as fixed stars unless there is another rationalization for the term *paʻa.* It is possible that *hōkūpaʻa* relates to the movement of these celestial bodies as having a fixed or established route. But that would not account for the *hōkū kiaʻi* that guide navigators being in the moving stars classification, since they follow an established routine as well and can be counted on to help navigators find their associated island groups. The rationale of Kepelino Keauokalani's stellar classification is elusive, and perhaps another generation of Hawaiʻi scholars will be able to find more clarity.

These physiological classifications of the islandscape are representative of how a Kanaka Hawaiʻi perspective describes a world full of intelligent bodies, each with their own vitality and responsibilities. Responsibility and sacrifice are very important concepts in the Kanaka Hawaiʻi perspective of living harmoniously with the multitude of our other-than-human relations. As Kanaka Hawaiʻi grew more populous, the need to manage their activities while maintaining the delicate balance of life cycles in the islandscape led them to create sociopolitical island classifications.

Island-centric sociopolitical landscape classifications

Most scholars who address these sociopolitical island classifications concentrate on either the organization of political leadership or the management of natural resources from a non-Hawaiʻi point of view. Beamer is one of many contemporary Hawaiʻi scholars contributing much-needed analysis and rigor for understanding Hawaiʻi history, leaders, and lands from a Kanaka Hawaiʻi point of view. While much of his work focuses on Hawaiʻi leadership during the colonial era of Hawaiʻi history, his research skills and fluency with the Hawaiʻi language allow him to provide the most concise presentations of the Hawaiʻi sociopolitical island classification system. He begins by discussing two concepts related to the responsibilities of the *mōʻī*, supreme leader (he suggests a possible explanation of "a succession of the supreme"): *palena* and *kālaiʻāina* (2014, 32–33).

Palena is a border, boundary, or dividing line between two places, but might also be explained as a protected place. Beamer describes *palena* as a place boundary created with a specific context that defines an area with unique functions (2014, 32–33). *Kālaiʻāina* is to manage or direct the human affairs of the land, and the person who engages in this activity is considered a *kālaiʻāina*. An important distinction must be made here: the *kālaiʻāina* did not manage the resources of the land; they managed the affairs of Kanaka Hawaiʻi as it pertained to the land. Managing resources comes from an anthropocentric tradition. Kanaka Hawaiʻi remembered their responsibility to maintain relationships and managed themselves, their conduct, and their character. So when the islandscape became filled with people, it was necessary to *kālaiʻāina* by imposing *palena ʻāina*, land boundary. This ensured that each area was nurtured in a manner that ensured longevity of the multitudes of peoples living within the imposed boundary. There are two types of *palena ʻāina*: large areas assigned to various levels of *aliʻi*, and smaller areas within them provided for *makaʻāinana*, general citizens or populace. Beamer focuses on the first type, as it directly relates to the administration and evolution of the Hawaiʻi government.

Fornander credits Māʻilikūkahi, an Oʻahu *mōʻī*, with the establishment of *palena ʻāina* on Oʻahu, because various cultivated and otherwise maintained lands of different sizes were in a state of confusion (1996, 89). Māʻilikūkahi divided the island into six *moku*, land areas that could be separated from one another with each standing on its own, and assigned each an *aliʻi nui ʻai moku*, a leader that eats from the district, where *ʻai*, to eat, is a metaphor for rule. Within these districts were sections of land known as *ahupuaʻa*, assigned to *aliʻi ʻai ahupuaʻa*, a leader that eats from the land area. An *ahupuaʻa* is a stone altar with a symbolic image of a pig set atop it. These altars were erected along a coastal path marking the boundary between land sections. It is where residents of the *ahupuaʻa* would place offerings as part of the *makahiki* celebration marking the start of the Hawaiʻi year.

Within the *ahupuaʻa* were three types of smaller land divisions known as *ʻili*: the *ʻili ʻāina*, the *ʻili kūpono*, and the *ʻili lele*. The *ʻili ʻāina*, also known as the *ʻili* of the *ahupuaʻa*, was assigned to warrior chiefs who shared directly with the *ali'i 'ai ahupua'a*. The *'ili kūpono* was assigned to *kaukau ali'i*, advisers, who shared their yields directly with the *ali'i nui 'ai moku*. According to Beamer, a special feature of the *'ili kūpono* is that it was not reassigned even if there were changes in leadership at the *ahupua'a* level. The *'ili lele* were several distinct land sections whose edges often did not touch each other yet formed one unit. Beamer does not indicate to whom yields from this type of *'ili* were given nor does he indicate what kind of *ali'i* was assigned to it. Mā'ilikūkahi also assigned lands within the ahupuaa to the *maka'āinana* so that they may cultivate relationships that provided yields they could consume themselves (Kamakau 1991, 54–55; Pukui and Elbert 2003; Beamer 2014).

The administrative boundaries set up by Mā'ilikūkahi have embedded within them a system of checks and balances. The larger land sections were assigned to *ali'i* of the highest rank and the smaller land sections were assigned to lower ranking *ali'i*, all of whom were loyal to the *mō'ī*. By having some smaller land sections share their yield with higher ranking *ali'i*, the *mō'ī* ensured any plans to usurp power or mismanage the sustainability of the land were discouraged. It behooved the *ali'i* to act appropriately in maintaining the balance between the various peoples living within the bounds of the assigned lands. Greed, ambition, or mistreatment of the people an *ali'i* was assigned to nurture resulted in dire consequences including death, war, replacement, or exodus. *Ali'i* may have been assigned to land areas, but *maka'āinana* were not, and though it was possible for *maka'āinana* to move from the lands of an unjust *ali'i* to the lands of a benevolent *ali'i*, the practicality of moving from a place where the bones of an ancestor rests makes it a rather rare occurrence. It is more likely that *maka'āinana* took it upon themselves to get rid of unjust *ali'i*, as the people of Ka'ū on Hawai'i Island are known to have done.

For the most part, the administrative boundaries set up by Mā'ilikūkahi were directly adopted by other island chiefdoms without change. However, Beamer's research provides documented evidence showing how two land terms, *kalana* and *'okana*, which have confused many scholars for decades, indicate this administrative system was not and should not be considered standardized or static. In an image of a page from what is commonly referred to as Buke Mahele,[11] Beamer shows how the term *kalana* is used to represent a land section similar to a *moku* for the island of Maui. He also found a newspaper article pertaining to the Kona side of Hawai'i Island nestled within the Hawaiian Ethnographic Notes (HEN) Index at the Bishop Museum Archives, translated by Mary Kawena Pukui, that indicates the term *ʻokana* is an additional land division "smaller than a moku division but larger than an ahupuaʻa" (2014, 40).

Hawaiʻi Island has hundreds of *ahupuaʻa*, more than any other island because it is larger than all the islands combined. As the population grew and *ahupuaʻa* became more numerous to accommodate this growth, it is conceivable that Hawaiʻi Island *aliʻi* added another level of administrative organization. What we do not know for sure is whether this new administrative level would have been a permanent addition or whether it was a temporary designation. It is possible that an *ʻokana* is a term that combines areas within a *moku* in order to split the *moku*, adding another *moku*. Nonetheless, the important point Beamer's research illuminates is there were variations to the *palena* administrative system that were contextually relevant to the places and situations each Island chiefdom faced.

The second type of *palena ʻāina* consists of smaller land sections within the *ahupuaʻa* provided to *makaʻāinana* for cultivation. They were named according to their appearance and function. For land areas useful for the production of wetland agriculture, there were terms such as *moʻo,* a long, narrow strip of land; *paukū,* a small lot of land; *kīhāpai,* a garden, field, or small farm; *kuakua,* a cultivated thin strip of land on the embankment between taro patches; and *loʻi,* a terraced irrigated pond. Dryland agriculture had terms such as *hakupaʻa,* a new taro patch; *mala,* a cultivated field for dryland planting; *ika,* the sides of a *mala* where grass is thrown; *iwi,* the side of a dryland taro field; and *malua,* a small hill dug up for planting potatoes (Kamakau 1976; Kent 1986; Pukui and Elbert 2003).

Also included in this group of *palena ʻāina* are two land areas, the *koʻele* and *hakuone.* The *koʻele* was cultivated by the *makaʻāinana* and assigned to the *konohiki*, a person who maintained the land and nearshore day-to-day activities of an *ahupuaʻa* and was loyal to the *aliʻi ʻai ahupuaʻa*. The *hakuone* was also cultivated by the *makaʻāinana* and assigned to the *haku,* a person who was the head of a household and loyal to the *ʻohana.* In a personal conversation, Andrade related that *makaʻāinana* were not required to maintain these land areas; they did it because oftentimes these leaders were busy ensuring the land continued to provide the necessary elements for their fields to be productive, such as free-flowing water. If a stream system became clogged with branches or fallen rocks, the leaders would clear it so the *makaʻāinana* could continue their daily activities. Working in the *koʻele* and *hakuone* ensured the leaders would have food even if they were too busy to tend the land themselves.

These "island as universe" classifications are not mutually exclusive. They overlapped when the human population grew large enough to require the establishment of leadership to ensure large-scale human activity remained within the valued principles of maintaining equitable balance with our older, other-than-human siblings. Thus, they are all intimately intertwined. Separating them from each other actually makes understanding each more difficult. For example, in a conversation with Kealiʻikanakaʻoleohaililani, she explained how a friend of hers wanted to save the native forest by removing the invasive species. She responded that if you pursued the same logic in the sociopolitical realm, there would be no nonnative peoples living here in Hawaiʻi. What a concept. Though to be clear, that was not necessarily the idea she was trying to get across to her friend. She explained further that, like any peoples, it is best to see whether indeed these so-called invasive species were detrimental to the overall function of the forest. Although she did not want to see her beloved Native relatives pushed out by aggressive nonnative peoples, the fact remains that it is already occurring at so many different layers of existence, and in many cases the Native relatives make the necessary adjustments to survive while providing space for the new peoples.

Kū i ka Honua (Pertaining to the Planet)

A Kanaka Hawaiʻi cartographic classification of the "earth as universe" involves understanding natural elements and processes, those divine entities many Kanaka Hawaiʻi consider the progenitors of all life on earth. In a previous chapter the phrase "divine entity" was given as an explanation for the Hawaiʻi language term *akua*. Many people relate to *akua* as gods, and those rituals practiced in honor of them are considered religion by some Hawaiʻi. This is unfortunate because it restricts the entire relationship Kanaka Hawaiʻi have with these divine entities by imposing non-Hawaiʻi social structures, wherein *akua* are revered from a distance and not experienced in the intimacy of the falling rain. Kanaka Hawaiʻi rapport with divine entities as familial extensions is a truly amazing feat. But what is even more astounding is the tacit acceptance that such a relationship is commonplace. In this section, I invite you to look beyond the rigidity of restrictive social structures and delve into the depth of the divine.

As Kanaka Hawaiʻi recognize literally hundreds if not thousands of divine entities, this section will present the major ones, based on occupancy and prevailing attitudes, as described by David Malo Kupihea in American folklorist and ethnographer Martha Beckwith's *Hawaiian Mythologies*: "According to [David Malo] Kupihea the great gods came at different times to Hawaii. Kū and Hina, male and female, were the earliest gods of his people. Kane and Kanaloa came to Hawaii about the time of Maui. Lono seems to have come last and his role to have been principally confined to the celebration of games" (1970, 11). Closer examination of each of these divine entities reveals a pattern of social and intellectual evolution.

The first seafaring peoples to begin populating the Islands of Hawaiʻi brought enough food and water to survive the voyage and plant foods and medicines. So, it is reasonable that they beseeched the divine companions Kū and Hina for assistance in the new lands because part of their *kuleana* is the fertility of the earth. Kū is associated with male fertility and that intangible essence in every entity to persevere, grow, adapt, and evolve. Hina is associated with female fertility and that innate nature in every entity to flourish, thrive, and multiply. Together they represent the duality of nature and participate in the production of food crops, good fishing, long life, and family. Furthermore, as *kū* is to rise, stand erect, and *hina* is to lean or fall from an upright position, when celestial bodies are in the eastern hemisphere they are associated with Kū and when they are in the western hemisphere they are associated with Hina.

Kū and Hina are divine entities men and women called on to conduct their daily tasks and, as such, had numerous manifestations. In some cases, Kū and Hina were considered complementary entities. For example, Kūʻulakai is the Kū entity associated with the abundant sea and revered by fishermen; however, if their requests for fish were not heeded, fishermen would then implore his companion, Hinapukuiʻa, the Hina entity from whom fish emerge, for assistance. Farther inland are Kūʻulauka, the Kū entity associated with the abundant inland, also known as Kū the cultivator, and Hinapukuʻai, the Hina entity from whom vegetable foods emerge. In the mountains are Kūkaʻōhiʻalaka, the Kū entity associated with the *ʻōhiʻa* tree, who is also connected with Laka, the entity associated with *hula*, and Hinaulu'ōhi'a, the Hina entity associated with the *'ōhi'a* forest growth (Beckwith 1970). These examples indicate early Polynesian settlers valued the complementary roles of these divine companions when cultivating the land. However, some Kū and Hina entities were implored for specific tasks, such as building a canoe.

Canoe builders would often spend part of their time in the forest to select, remove branches from, and haul large trees as part of the canoe-building process. They recognized Kūmokuhāli'i, Kū entity associated with the spreading land, to support the emigration or spreading of people to new lands; Kūpā'aike'e, Kū entity associated with the stone-devouring bent-shaped adze, to maintain the integrity of the trusted adze crafted and honed to bring down large trees; Kūpulupulu, Kū entity associated with the tinder or kindling maker, to encourage the ease of removing excess bulk, reducing the weight prior to hauling the tree downhill; and Kūholoholopali, Kū of the steep haul downhill, to help protect the tree when it is in transit being hauled down steep places (Beckwith 1970, 15–16). Each entity has its specific *kuleana*, and men dutifully quieted their minds and focused their intent in reverential silence, communicating with ever-present entities who granted or withheld their favor to the requests. Success meant focused energy remained concomitant among all entities involved, and failure was a reason to reassess practices, eliminating distractions and perfecting procedure—a process similar in some aspects to the scientific method. Indeed, embedded within even these early arriving entities were the elements of a Kanaka Hawai'i understanding of science.

A Kanaka Hawai'i scientific understanding includes recognizing that each element and process in the natural world has specific *kuleana*, such as Kāne and Kanaloa, the next divine companions to arrive in Hawai'i according to Malo. Their most commonly known *kuleana* are the waters of the earth, both fresh and salty. Kāne is the entity associated with life-giving and sustaining natural elements and processes such as the sun, air, clouds, moisture, mist, and rain. Kanaloa is the entity most commonly associated with the elements and processes of the ocean, such as specific currents, surf conditions, wave actions, and marine organisms. Papahulihonua practitioner Ku'ulei Higashi Kanahele adds that Kanaloa is also associated with particular land elements and processes such as freshwater aquifers, rift zones, pumice, and estuaries (2014). Keali'ikanaka'oleohaililani revealed in a conversation that many Kanaka Hawai'i perceive a world where interaction with divine entities is an everyday occurrence, stating, "When we breathe we embody *Kāne*, when we swim in the ocean, we return to the bosom of *Kanaloa*" (2013). These intimate interactions and observations of the natural world were recorded in several stories, many of which were condensed into chants in a manner that is still contextually relevant today. For example those processes that are part of the hydrologic cycle are described in a well-known chant, shared by Emerson:

"Wai o ke Ola"
(Waters of Life)

He ui, he nīnau
A query, a question
E ui aku ana au iā ʻo
I put to you
Aia i hea ka wai a Kāne?
Where is the water of Kāne?
Aia i ka hikina a ka Lā
At the Eastern Gate
Puka i Haehae
Where the sun comes in at Haehae
Aia i laila ka Wai a Kāne
There is the water of Kāne

E ui aku ana au iā ʻoe
A question I ask of you
Aia i hea ka wai a Kāne?
Where is the water of Kāne?
Aia i kau lana ka lā
Out there with the floating Sun
I ka pae ʻōpua i ke kai
Where cloud forms rest on Ocean's breast
Ea mai ana ma Nihoa
Uplifting their forms at Nihoa
Ma ka mole mai o Lehua;
The side the base of Lehua
Aia i laila ka Wai a Kāne
There is the water of Kāne

E ui aku ana au iā ʻoe
A question I ask of you

Aia i hea ka wai a Kāne?
Where is the water of Kāne?
Aia i ke kuahiwi, i ke kualono,
On mountain peak on the ridges steep
I ke awāwa, i ke kahawai;
In the valleys deep, where rivers sweep
Aia i laila ka Wai a Kāne
There is the water of Kāne

E ui aku ana au iā ʻoe
A question I ask of you
Aia i hea ka wai a Kāne?
Where is the water of Kāne?
Aia i kai, i ka moana,
At sea, on the ocean
I ke Kualau, i ke ānuenue,
In the driving rain, in the heavenly bow
I ka pūnohu, i ka uakoko,
In the piled up wraith, in the red rainfall
I ka ʻālewalewa;
In the ghost pale cloud form
Aia i laila ka Wai a Kāne
There is the water of Kāne

E ui aku ana au iā ʻoe
A question I ask of you
Aia i hea ka wai a Kāne?
Where is the water of Kāne?
Aia i luna ka Wai a Kāne,
Up on high is the water of Kāne
I ke ʻōuli, i ke ao ʻeleʻele,
In the portent, in the black piled cloud
I ke ao panopano,
In the black, black cloud
I ke ao pōpolohua mea a Kāne la, e!
In the black mottled sacred cloud of Kāne
Aia i laila ka Wai a Kāne
There is the water of Kāne

E ui aku ana au iā ʻoe
A question I ask of you
Aia i hea ka wai a Kāne?

Where is the water of Kāne?
Aia i lalo, i ka honua, i ka wai hū,
Deep in the ground, in the gushing spring
I ka wai kau a Kāne me Kanaloa
In the ducts of Kāne and Kanaloa
He waipuna, he wai e inu,
A well-spring of water, to quaff
He wai e mana, he wai e ola.
A water of magic power, the water of life
E ola no, e a!
Life! Give us life!
(Emerson 1998, 257–259)

Many people refer to this chant as depicting the importance of water to Kanaka Hawaiʻi. Oftentimes, Kanaka Hawaiʻi nationalists evoke this chant in protest as they continue to question the diversion of freshwater from their natural routes for the purposes of development. My metaphoric mind recognizes how this chant also demonstrates and maintains a Kanaka Hawaiʻi understanding of the hydrologic cycle in the Islands of Hawaiʻi.

In the first stanza, the composer recognizes the sun as a necessary entity in the hydrologic cycle. *Haehae*, to tear or rip, is a metaphorical portal, the eastern gate of heaven, through which the sun enters, or tears into, this reality. The sun is important for evaporation or the way water changes form from a solid into a vapor. The second stanza describes condensation or the formation of clouds, relating this process back to the sun. *Kau lana ka lā*, the floating sun, is the area of the heavens where the sun seems to be drawn out in a boatlike form, appearing to float on the surface of the ocean. The phrase became identified with the locality of the phenomenon and thus is considered a proper name. *Pae ʻōpua i ke kai,* cumulus or billowy, puffy cloud banked up near the horizon, is another instance of a natural phenomenon being given a proper name.

The third stanza describes precipitation and surface runoff. The composer observed how clouds rising over the mountains cause precipitation in a process known as the orographic effect. As the warm air from the ocean rises to climb the mountains, it cools, condenses, and falls to the ground creating *kualono, awāwa*, and *kahawai*. The fourth stanza describes how rain originating from the ocean warns of approaching storms. It is similar to the popular mariner's phrase, "Red sky in the morning, sailors' warning." Many Kanaka Hawaiʻi recognize how a reddish sunrise, depicted by *kualau*, wind-driven rain, and *uakoko*, red rainfall, often foreshadows an approaching storm.

The fifth stanza describes the different types of clouds that carry water—from *ʻeleʻele*, the darkened clouds piled high, to *panopano*, the thick black clouds, and *pōpolohua*, the purplish-blue reddish-brown clouds, which are considered special to Kāne. The last stanza describes infiltration and subterranean water sources. It identifies water as being *i lalo, i ka honua*, deep in the ground, *i ka wai hū*, gushing up from springs, and contained in *i ka wai kau*, aqueducts. *Wai kau a Kāne me Kanaloa* is an acknowledgment of a relevant story of Kāne and Kanaloa from Maui. Once when they were traveling together, Kanaloa became thirsty and Kāne thrust his staff into the *pali* at Keʻanae, Maui, and out flowed a stream of pure water that has continued to the present day. The reason both Kāne and Kanaloa are acknowledged in the process of bringing springwater to the land surface is because both are necessary components. Kanaloa is the large, deep, immovable aquifer from which the life-giving water of Kāne originates (K. H. Kanahele 2014). An important point to recognize with this chant is that it is as accurate today as it was the day it was composed. It is not some relic from days long past. It is a methodically observed, symbolically encoded, oral record of scientific processes.

Kanaka Hawaiʻi also associated Kāne and Kanaloa with the movement of the sun, as shown in the previous chapter. Kāne and Kanaloa were essential entities that demarcate the sun's farthest northward progression during the summer solstice, *ke alanui polohiwa a Kāne,* the large dark glistening path of Kāne, and the sun's farthest southward progression during the winter solstice, *ke alanui polohiwa a Kanaloa,* the large dark glistening path of Kanaloa. Interestingly, although the east is associated with Kū because it is where celestial bodies appear to rise from the horizon, and the west is associated with Hina because it is where celestial bodies appear to fall into the horizon, the arrival of Kāne and Kanaloa added another layer of celestial understanding. The eastern sky became known as *ke ala ula a Kāne,* the flaming path of Kāne, and the western sky became known as *ke alanui maʻawe a Kanaloa,* the great wispy path of Kanaloa. This indicates an evolved understanding of the movement of celestial bodies. Not only did Kanaka Hawaiʻi recognize celestial bodies appear to rise and fall in the east and west, respectively. They also observed the north–south movement of the sun, built large-scale structures to track that movement, and recorded their understanding in the names of the celestial phenomenon.

According to Malo, Māui arrived in Hawaiʻi about the same time as Kāne and Kanaloa. However, Maui is not considered a major entity, as he is not associated with specific natural elements or processes. Nonetheless, Maui is an important ancestor, whose activities are woven into the fabric of Kanaka Hawaiʻi consciousness as acts of heroism, mischief, and self-assurance. Similar stories of Maui are found all over the Pacific, with three specific nodes of origin in New Zealand, Tahiti, and Hawaiʻi. The overwhelming similarity in almost all stories of Maui is that he represents the combination of intellectual inquiry, ingenuity, and action. Māui identifies a problem, plans a strategy, and acts on it without hesitation and, in some cases, adjusts his actions such that he achieves his goal. Again, the scientific method echoes in some ways with this storyline, because the projects Māui undertakes transform the way people live and work in their island homes.

Lono is the last of the great divine entities identified by Malo but certainly not the last to arrive in the islands. As mentioned in previous chapters, Lono is the entity associated with agricultural fertility, peace, and recreation. He emerges in stories and chants when the Hawaiʻi population is burgeoning and has established a well-defined class system with leadership and military strength. Lono offers the masses the guarantee of peace from war and respite from large-scale laborious tasks. Meanwhile the leadership takes the opportunity to recuperate from battle, refill their reserves, and hone their warriors with games of endurance, speed, and agility. A large population necessitates an abundance of food, and Kanaka Hawaiʻi in this epoch were such ingenious and prolific farmers that they made even the driest parts of the islands fertile, developing innovative agricultural systems that were both self-sufficient and sustainable.

Lono is also associated with the wet season of *hoʻoilo*, when dark rain clouds and winter storms are welcomed by the peoples on the dry, leeward sides of the islands as they help to keep the dry lands fertile. Where Kū and Hina are responsible for the fertility of the earth, Lono is that atmospheric entity that ensures agricultural productivity. Many of the natural phenomena associated with *hoʻoilo* are aspects of Lono, such as thunder, lightning, whirlwinds, waterspouts, gushing springs, and red-stained streams. Lonomākua, a Lono entity who presides over *makahiki*, is also associated with being the fire keeper, a revered relative of Pele, who is a more recent divine entity to arrive in Hawaiʻi.

Knowledge of Pele is prevalent throughout the islands, as her escapades from Kauaʻi to Hawaiʻi have been shared by storytellers, chanters, and dancers for generations. Today, many Kanaka Hawaiʻi understand that Pele is a person as well as an entity associated with volcanism. As a person, Pele arrived with her clan and originally resided on Kauaʻi. Eventually the clan moved through the islands and settled on Hawaiʻi, but by the time they arrived on the Island of Hawaiʻi, Pele had transitioned into the entity intimately associated with the natural elements and processes of volcanism. A native of Kaʻū, Pukui articulately describes the Hawaiʻi connection with Pele as a beloved divine familial entity:

> If Pele is not real to you, you cannot comprehend the quality of relationship that exists between persons related to and through Pele, and of these persons to the land and phenomena not "created by" but *which are*, Pele and her clan. A rosy dawn is not merely a lovely "natural phenomenon": it is that beloved Person named "The-rosy-glow-of-the-Heavens," who is "Hiʻiaka-in-the-bosom-of-Pele," the youngest and most beloved sister of that greater (and loved though awe-inspiring) Person, Pele-honua-mea (Pele-the sacred-earth-person), whose passions express themselves in the upheavals of vulcanism, whose "family" or "clan" are the terrestrial and meteorological phenomena related to vulcanism and the land created by vulcanism, as actively known in Ka-ʻu. (Handy and Pukui 2010, 28; italics, capitalization, and dashes original)

The Pele peoples have maintained knowledge of the "terrestrial and meteorological phenomenon related to volcanism" for generations. In her most recent publication, *Ka Honua Ola: ʻEliʻeli Kau Mai (The Living Earth: Descend, Deepen the Revelation)*, Kanahele shares her explanations and interpretations of the Pele clan that she has spent a lifetime unwrapping, providing points of access to a larger audience (2011). Pele has many genealogies depending on where and from whom the composition originates. In the Pele family genealogy found in *Ka Hoku o Hawaii*, a nineteenth-century Hawaiʻi language newspaper based in Hilo, Kanahele expresses her insights into how Pele peoples maintained scientific knowledge associated with volcanism.

"Haumea Lāua ʻo Moemoeaʻaliʻi"
(Haumea Together with Moemoeaʻaliʻi)

Makuahine: Haumea
Mother: Haumea
Makuakane: Moemoeaʻaliʻi
Father: Moemoeaʻaliʻi
Nā keiki a lāua, ke one hānau

Their children and place of birth
Kamohoaliʻi, hānau ma ka manawa mai
Kamohoaliʻi, born from the fontanel
Kānehekili, hānau ma ka waha
Kānehekili, born from the mouth
Kauilanuimākēhāikalani, hānau ma ka maka
Kauilanuimākēhāikalani, born from the eyes
Kūhaʻimoana, hānau ma ka pepeiao
Kūhaʻimoana, born from the ears
Kānemilohaʻi, hānau ma ka lima ʻākau
Kānemilohaʻi, born from the right palm
*Leho, hānau ma ka ʻ*ōpu*ʻupuʻu lima*
Leho, born from the knuckles
Kāneikōkala, hānau ma ka manamana lima
Kāneikōkala, born from the fingers
Nāmakaokahaʻi, hānau ma ka umauma
Nāmakaokahaʻi, born from the chest
Pelehonuamea, hānau ma kahi mau e hānau ʻia ai ke kanaka
Pelehonuamea, born from the usual place of people
Kapōʻulakīnaʻu, hānau ma nā kuli
Kapōʻulakīnaʻu, born from the knees
*Kapōkohelele, hānau ma nā ʻ*ōpu*ʻupuʻu wāwai*
Kapōkohelele, born from the ankles
Hiʻiakakalukalu, hānau ma nā manamana wāwai
Hiʻiakakalukalu, born from the toes
Hiʻiakakuilei, hānau ma nā kapuaʻi wāwai
Hiʻiakakuilei, born from the feet
Hiʻiakaikapoliopele, hānau ma nā poho lima
Hiʻiakaikapoliopele, born in the palms
ma ke ʻano me hehua moa ala
in the shape of an egg
(2011, 2–3)

Haumea is the entity associated with life-forming processes on land and in the sea. Moemoeaʻaliʻi is the entity associated with dormancy or the latency and potentiality of life. The composition reveals that the union of these two entities produces fourteen offspring. The names and birthplaces of each offspring provide clues about their form and function in the volcanic construction of the Islands of Hawaiʻi. Kamohoaliʻi, the firstborn, is the entity associated with locating potential sites for volcanic activity to occur.

Kamohoaliʻi's birthplace, the fontanel, is associated with those soft spots on the crust of the earth from which magma can surface. Kamohoaliʻi often takes the form of a shark, as sometimes those soft spots occur underwater, especially in the Pacific.

Kānehekili, the second born, is the entity associated with the thunder that rumbles from within the earth. Kānehekili's birthplace is the mouth and is associated with the sounds emanating from the movement of magma through the earth. Kauilanuimākēhāikalani, the third born, is the entity associated with lightning that originates from the earth, not the sky. Kauilanuimākēhāikalani's birthplace is the eyes and is associated with the electrical energy that surges upward from the ground where magma moves within the earth. Kūhaʻimoana, the fourth born, is the entity associated with lava tubes that span the ocean floor. Kūhaʻimoana's birthplace is the ear. The ear canal both detects the thunderous sounds of large quantities of moving magma and appears to be physically similar to horizontal lava tubes that span the ocean floor. Kānemilohaʻi, the fifth born, is the entity associated with lava tubes that originate from deep in the earth. Kānemilohaʻi's birthplace is the right palm or, more specifically, the hollow part of the cupped palm of the right hand, which is associated with the vertical movement of large quantities of magma (Kanahele 2011, 5–6).

Kanahele does not provide insight on either Leho, the sixth-born entity, originating from the knuckles, or Kāneikōkala, the seventh born, originating from the fingers. The following is my attempt to understand these entities in relation to volcanic processes. *Leho* is a callus or a thickened, hardened part of the skin that has been subjected to repeated pressure, friction, or irritation. When calluses form on the skin they often look discolored, much like sulfur banks. It is my belief that Leho is the entity associated with processes that create sulfur-crusted earth. Kuʻulei Higashi Kanahele describes Kāneikōkala as the entity associated with volcanic dykes, those vertical veins of magma that cut through older rocks (2015).

Nāmakaokahaʻi, the eighth born, is the entity associated with geologic faults or weak points in the earth that easily break away and allow the magma to reach the surface. Nāmakaokahaʻi's birthplace, the chest, is associated with the broad flat plane of land where these geologic faults occur (Kanahele 2011, 6). Pelehonuamea, or Pele, is the ninth born and is the entity associated with lava and the creation of land. Pele is born from the usual place correlating a volcanic crater with a birth canal. Pualani Kanahele does not share her explanation or interpretation on the next three offspring, but Kuʻulei Higashi Kanahele does provide some insights.

Kuʻulei Higashi Kanahele recognizes all the offspring born before Pele as those processes that occur before an eruption and those offspring born after Pele as those processes that happen after an eruption. Thus, Kapōʻulakīnaʻu, the tenth offspring, is the entity associated with the direction lava will take on the surface of the earth. Born from the knees, Kapōʻulakīnaʻu sits in judgment, determining the route lava will take. Kapōkohelele, the eleventh offspring, is born from the ankles and is associated with the spreading of lava. Hiʻiakakalukalu, the twelfth offspring, is born from the toes and is associated with the thinning of lava as it spreads across the surface of the earth (Kanahele 2015).

Pualani Kanahele indicates Hiʻiakakuilei, the thirteenth offspring, born from the feet, is the entity associated with the alignment of cones, mountains, and islands, much like footsteps in the sand, and represents the continuum of volcanic activity. Kuʻulei Higashi Kanahele believes Hiʻiakakuilei may also have something to do with the dispersal of seeds needed for new growth. Both agree that Hiʻiakaikapoliopele, the fourteenth and last offspring, born from the palms in the shape of an egg, is the entity associated with generating life on the barren lava fields (K. H. Kanahele 2015; P. K. Kanahele 2011, 6–7).

Some of these offspring appear in other *mele,* Hawaiʻi song, chant, poem, and in *moʻolelo*, historical accounts, two forms of knowledge communication typically categorized as folklore or myth; sadly, this means the many levels of metaphoric understanding embedded within them, including the advanced scientific observations shown here, are easily overlooked. Kanaka Hawaiʻi, like many peoples worldwide, possess profound scientific wisdom. Preconceptions of the package these scientific observations are presented in do not make the information any less relevant or more important. These knowledge packages have served Kanaka Hawaiʻi well for many generations as they fertilize the metaphoric mind, ensuring the sensual relationship maintained with our other-than-human entities continues to flourish. Thankfully, Hawaiʻi scholars in various disciplines who have learned the intricacies of Hawaiʻi language and have passionately engaged in a range of Hawaiʻi cultural practices that further develop their metaphoric capabilities are now uncovering the depth of the divine wisdom these texts have maintained. These scholars enlighten their disciplines with (k)new[12] knowledge from the ancestral depths embedded in these recorded texts, knowledge that is as evident today as it was when these sources were created centuries ago.

Understanding life as cyclical, seasonal, and evolutionary is as simple as looking up into *kalanipaʻa* and watching the movement of the celestial bodies. Recognizing the connection between the celestial bodies and their effects on the island environment is as easy as going to the shoreline and watching how the ebbing tide reveals a reef teeming with life. Embracing natural elements and processes as entities with responsibilities with whom human peoples can daily relate and communicate is more common than most people want to admit. The truly difficult part of embracing Kanaka Hawaiʻi scientific understandings is trusting one's metaphoric mind to understand patterns beyond boundaries. Yet it is the simplicity of recognizing nature as self that breaks through the monotony, tedium, and rigidity of neatly categorized and separately studied knowledge systems.

It is important to remember that Kanaka Hawaiʻi spatial/temporal knowledge is much more fluid than presented in this chapter. Concepts were organized herein to best represent a Kanaka Hawaiʻi cartographic knowledge system. Another scholar from a different discipline with a different focus or with a different set of experiences can take the same information presented here and organize it completely differently. This does not make the knowledge, organization of knowledge, or presentation of knowledge any less or more genuine. It simply means Kanaka Hawaiʻi knowledge systems are both culturally constructed and contextually relevant.

PART 2

Nā Hana Hawai'i

(Hawai'i Practices)

Kanaka Hawai'i cartographic practices are shaped by distinct philosophical underpinnings and steeped in contextually relevant spatial/temporal knowledge perspectives and classifications. The chapters in part 2 build on the Kanaka Hawai'i cartographic foundations described in part 1, providing further understanding of a sensually embodied, encoded, and embedded Kanaka Hawai'i spatial/temporal knowing. In order for Kanaka Hawai'i cartography to achieve solutions to relational problems, it must engage in three essential processes: spatial/temporal knowledge acquisition, representation, and transmission.

Ka ho'okele is the first Kanaka Hawai'i cartographic practice presented because successful voyages require a sensual awakening. The navigator must allow all his/her senses to alert him/her to changes in the environment that can lead to either fortuitous or dangerous conditions. *Ka ho'okele* requires more than just knowing about the stars; it also requires feeling the directional movement of the ocean, smelling the distant storm, and listening to the voices of the ancestors through the fabric of space/time. It is an opportunity to understand alternative modes of acquiring spatial/temporal knowledge.

Ka haku 'ana provides an opportunity to understand how Kanaka Hawai'i spatial/temporal knowledge is encoded or represented, because it requires sensual awareness. Almost every form of Kanaka Hawai'i oratory uses specific techniques with symbolic meaning and includes *wahi pana,* celebrated places with storied significance, from our *mele inoa,* name song composed to honor a person or place, to our *kanikau,* lament or song of mourning. These compositions are elegantly arranged using several sensually oriented oratory techniques, including timing, rhythm, and pacing, to create a connection to another space/time where/when the topic of the oratory occurs.

A great composer does this so effortlessly that the observer/participant is unaware of how sensually stimulated and emotionally invested they are in the oratory.

Ka hula highlights the way Kanaka Hawaiʻi transmit spatial/temporal knowledge because it requires a sensual attentiveness to the integration of the human mind/body with nature/culture. *Hula* is also an integration of several cultural practices. Successful performances require instruments, clothing, adornments, and vocal and dance training. *Nā kumu hula,* Hawaiʻi dance teachers, embed multisensual symbolic nuances that effectively change the apparent stability of space/time, arousing the mind/body of the viewer/participant to be transported into another realm.

For those readers still struggling to embrace the Kanaka Hawaiʻi cartographic consciousness presented thus far, it is time to rip the bandage, remove the blinders, or take the plunge. Embracing a Kanaka Hawaiʻi cartographic consciousness means being outside a comfort zone that has protected mainstream knowledge traditions from evolving beyond the confines of the scientific method. The time is ripe for another Kuhnian revolution[13] of global consequence, where the perspective and direction of scientific inquiry leads to a more inclusive, multicultural understanding of reality as a pluriverse.

Here are some helpful hints I learned from Kealiʻikanakaʻoleohaililani that will assist those still struggling to be able to understand Hawaiʻi chant texts such that their metaphoric meanings and Kanaka Hawaiʻi cartographic relevance are not lost. Discard anything you thought you knew about any of Hawaiʻi *akua,* divine entities, such as Pele, popularly known as a legendary *akua* of the volcano. Consider *akua* as any energy, process, or entity that is required for any being to live, sustain life, and procreate. Create a knowledge construct of *akua* that elevates its physical aspects and ecological contributions to being equal with its spiritual significance. This means you cannot focus on *akua* as god or compartmentalize *akua* as a religious entity in a Western construct. Lastly, recognize Hawaiʻi chant texts as compact mobile records of long-term observations of the natural world, putting them on par with any other source of knowledge (Kealiʻikanakaʻoleohaililani 2013).

CHAPTER FOUR

Ka Hoʻokele

(Hawaiʻi Navigation)

"Mai Kahiki ka Wahine ʻo Pele"
(From Kahiki came the Female, Pele)

Mai Kahiki ka wahine ʻo Pele
From Kahiki came the female, Pele
Mai ka ʻāina i Pola-pola
From the land of Borabora
Mai ka pūnohu ʻula a Kāne
From the red rising mist of Kāne
Mai ka ao lapalapa i ka lani
From the agitating clouds in the sky
Mai ka ʻōpua lapa i Kahiki
From the churning clouds of Kahiki
Lapakū i Hawaiʻi ka wahine, ʻo Pele
The woman Pele explodes to Hawaiʻi
Kālai i ka waʻa Honuaiākea
The vessel Honuaiākea is carved
Kou waʻa, e Kamohoaliʻi
It is your vessel, Kamohoaliʻi
I ʻapoa ka moku i paʻa
The island will be gotten and secured
Ua hoa ka waʻa o ke akua
The vessel of the god is completed
Ka waʻa o Kānekālaihonua
The vessel of Kānekālaihonua
(Kanahele 2011, 36–37)

All Kanaka ʻŌiwi Hawaiʻi, Native persons of Hawaiʻi genealogical lineage, descend from people who crossed the Pacific Ocean in a *waʻa kaulua*, long-distance double-hulled voyaging canoe, arriving in the Islands of

Hawaiʻi as early as 1000 CE.[14] Such feats of exploration are no trivial matter. The Pacific Ocean occupies nearly one-third of the planet's surface area. Dotted within it are approximately twenty-five thousand islands, some with more than two thousand miles between them. The crafts used for voyaging were hand-carved with tools made by hand from natural materials such as stones, bones, and corals; rigged with handwoven fibers; and navigated without the use of instruments (Henry and Kawaharada 1995). The immensity of this accomplishment cannot be overstated, or even truly understood in modern times, yet a thousand years ago one of my ancestors took that daring voyage and ventured beyond familiar coastlines. While I may not know the specifics of that voyage, a few voyages of prominent ancestors were recorded in mele, Hawaiʻi song, chant, or poem, as presented in the opening quote.

"Mai Kahiki ka Wahine ʻo Pele" is a chant from the *hula Pele* repertoire, a Hawaiʻi dance genre that honors Pele. The portion of the chant shared in the opening quote indicates that long before Pele became associated with volcanism, she was an essential member of a voyaging family. According to Kanahele, the first few lines of this chant indicate Pele is mindful of the protocol of asking permission before entering the realm of another entity and adheres to those distinctive protocols that are cultural imperatives for a successful voyage (2011, 38). Recognizing relationships and asking permission for a safe and successful undertaking is a matter of Hawaiʻi social etiquette that is blind to profession, rank, or power and extends to all entities, not just the human entities. Proceeding without attending to such etiquette is not only considered rude—it also comes with consequences.

This chant suggests that Pele asked Kāne for permission to travel to the Islands of Hawaiʻi. In one of Pele's genealogies, Kānehoalani, the Kāne entity associated with the life-giving processes of the sun, is Pele's father. According to Kealiʻikanakaʻoleohaililani, Pele people recognize the relationship between the sun, Kānehoalani, which they consider to be the ultimate volcano, and volcanic activity on earth. So it was no surprise to them when scientists correlated solar flares with increased seismic activity on earth. Forming relationships and communicating with natural entities is common for some Kanaka Hawaiʻi, as is understanding the responses those natural entities offer. Kanahele says the entities named in the lines *ka pūnohu ʻula a Kāne*, the red rising mist of Kāne; *ka ao lapalapa i ka lani*, the agitating clouds in the sky; and *ka ʻōpua lapa i Kahiki*, the churning clouds of Kahiki, are atmospheric entities associated with Kāne and indicate Pele's request to travel has been granted.

There are several variations to the stories of why Pele left Kahiki for Hawaiʻi. One of the first ones I heard as a child involved a tryst with the *kāne,* male, man, husband, of her sister, Namakaokahaʻi, who became so enraged she chased Pele away. In *Unwritten Literature of Hawaiʻi,* Nathaniel Bright Emerson, medical physician and self-trained Hawaiʻi ethnographer, explained that Pele's brothers expelled her from their home because she disrespected their mother, Haumea, by scorching the earth (1998, 188). Kanahele, however, provides an alternate version. She describes the voyage as one with a known purpose and destination. Pele intended to travel first to Polapola, then to the islands of Kūʻaihelani and Manamana (both now atolls in the Northwestern Hawaiʻi Islands), and the place of the *aliʻi wahine,* Kaʻōahi (the island of Niʻihau) (2011, 38). I consider all variations of this story correct. The first story maintains an evangelically preferred societal morality, the second describes a physical geological occurrence, and the last explains Pele's journey as a pilgrimage. I believe it is precisely because Hawaiʻi *moʻolelo,* Hawaiʻi historical accounts, are able to adapt to societal shifts in power and influence that they are able to survive, allowing the deep knowledge embedded within them to be accessed by masters such as Kanahele and Kealiʻikanakaʻoleohaililani.

The last few lines of the opening chant suggest Pele surges forth to the Islands of Hawaiʻi on two vessels carved specifically for the voyage. One is named *Honuaiākea,* the expanse of the earth, for her eldest sibling Kamohoaliʻi, and the other is named *Kānekālaihonua,* the Kāne entity that carves the earth. Together these two canoes make up the substance and function of the Pele family (Kanahele 2011, 38). These lines reveal that the journey was planned and purposeful, as Kanaka Hawaiʻi cartographic practices such as *hoʻokele* require years of preparation. Oftentimes the observer/participant has little to no idea what is required of practitioners to succeed in their craft. Though the amount of time and sacrifice each practitioner fulfills is specific to their practice, examining some of the preparations necessary for a successful long-distance voyage will allow us, as the observer/participant, to be more invested in the Kanaka Hawaiʻi cartographic elements of *ka hoʻokele.*

Navigation is the act of executing a planned route from a known origin to a known destination. *Hoʻokele* is a distinct form of non-instrument navigation replete with Kanaka Hawaiʻi cartographic processes. It uses body-centric measurements, island-centric skyscape classifications, and planet-centric understandings of astronomical cycles and atmospheric movements. The focus here is on awakening a sensual consciousness through the embodiment of spatial/temporal knowledge. Many seafarers understand how successful journeys across an ocean expanse such as the Pacific require rhythmic synchronicity with the vessel, the crew, and the elements of nature. For some people, slipping into a hypnotic blend of movement and adapting to the rhythms of the natural world is like putting on a well-worn pair of jeans, as their sensual consciousness is already awake and attuned to the elements and processes of nature. We begin explaining the preparations necessary for a long-distance non-instrument ocean voyage before describing the embodied sensual awakening that occurs as part of a voyage. It is important to note that *hoʻokele* is not a practice everyone can undertake, as the author learned early in life. As a result, much of the information shared in this chapter is derived from texts and not from personal experience.

Preparing the Navigator

For Kanaka Hawaiʻi, long-distance voyages are a purposeful, mindfully focused, and prayer-filled undertaking. Preparation for a long-distance voyage is neither a quick and easy task nor is it completed without careful attention to protocol. It requires a *waʻa kaulua*, a well-trained crew, and enough provisions to sustain the voyage. Each of these requirements involves careful long-term planning and commitment. For example, building canoes was more than an important profession requiring years of training; it was an industry involving entire communities and several practitioners. The *kahuna kālaiwaʻa*, master canoe carver, was the technical and spiritual adviser who made all major decisions in the canoe-building process, praying at every stage, from selecting a tree, felling it, hewing it, and hauling it to finishing it and launching it. The entire process took several months, as the *kahuna kālaiwaʻa* ensured each step was attentively completed and each canoe was imbued with *mana* through appropriate ceremony (Kamakau 1976, 119; Pukui et al. 1972, 131). Yes, a canoe is considered a living being with purpose, responsibility, and intelligence and with whom the navigator and crew form a lifelong and timeless bond, as each cares for the safety of the other.

A competent, properly trained navigator is one of the most important elements of any successful long-distance ocean voyage, as the navigator is responsible for guiding the vessel and its crew to its destination. Navigating any ocean vessel without modern instruments is nothing like driving a car without a GPS, even though inattention to operating either vehicle can be disastrous if not deadly. A car is not at the mercy of natural forces such as wind and waves to move it and the stars and clouds to guide it and the animals, feathered and finned, to indicate potential rest stops for it. A Hawai'i navigator must be trained for several years before being deemed ready to take the helm. *Ho'okele* involves more than just the knowledge of the stars, swells, and winds; it requires a person capable of accessing the depths of ancestral knowledge, where consciousness shares space/time with the subconscious and collective consciousness. Knowledge of the stars, swells, and winds was imparted to hopeful youths who were selected at an early age, all of whom demonstrated characteristics or tendencies a master navigator identified as necessary for his trade. Star knowledge was one of the first things imparted, as intimacy with the locations and movement of stars increases accuracy of position detection at sea when there is no other frame of reference, especially on long-distance non-instrument voyages.

Hawai'i political scientist Noenoe Silva located texts about the practice of *kilo hōkū,* astronomy, written by Hawai'i writer, editor, attorney, and politician Joseph Mokuohai Poepoe, titled *Moolelo Hawaii Kahiko* (Ancient Hawai'i History). This nine-month series of articles was published in *Ka Na'i Aupuni*, a Hawai'i language newspaper edited by Poepoe. Silva notes that Poepoe divided the section titled "Ka Oihana Kilo Hoku Hawaii," Hawai'i astronomy, into three subsections: "1) Ke kulana kahiko . . . o ka Oihana Kilo Hoku ma Hawaii nei, a description of Hawaiian astronomy; 2) Ke Ao ana i ka Oihana Kilo Hoku, how astronomy was taught; and 3) Na Hoakaka no na hoku i keia maloko o keia oihana, explanation of the stars that were known in this branch of knowledge" (Silva 2017). Since much of the information Poepoe shares in sections 1 and 3 has been described in part 1 of this text, the focus here is on section 2. Poepoe begins section 2 by stating, "E like me ke a'o ana o ke kanaka i ka ike hula, ike lapaau, a pela aku, i ke au kahiko, ma ke ano kapu a ihiihi, peia no ke a'o ana i ka oihana kilo hoku Hawaii" (Learning Hawai'i astronomy is similar to learning Hawai'i dance or Hawai'i medicine with regard to the sacredness and reverence) (Poepoe 1906).

Navigational acolytes knew they were required to demonstrate their mastery of the knowledge at each step before they could advance to the next step. It was a scaffolding teaching technique. They started by memorizing the names of stars, then orating them flawlessly and without prompting, before moving on to the next step. After mastering the names of stars, they learned *pule*, prayer, about the stars. After flawless mastering recitation of *pule,* they would sit with a *kumu*, Hawai'i teacher, on a square mat plaited with finger-sized mesh strips, where the *kumu* would arrange pebbles into star clusters beginning with Makali'i, Seven Sisters or the Pleiades, and moving to other named star groups. Once the student mastered the star arrangements, they learned the order, the location, and the times of night that each of the star groups appear to rise and set along the horizon (Poepoe 1906).

Poepoe uses the majority of this section to describe a specific *ipu hōkeo,* long gourd bowl, that had many lines and dots burned into it. The *kumu* used this device to help students learn the different parts of the *ka lani pa'a*, the celestial sphere, such as *ke ala ula a Kāne*, eastern sky; *ke alanui ma'awe a Kanaloa*, western sky; *ke alanui polohiwa a Kāne*, celestial path of the sun at summer solstice; *ke alanui polohiwa a Kanaloa,* celestial path of the sun at winter solstice; and *ke alanui i ka piko o Wākea*, celestial path of the sun at the equinox. At the head of the bowl rested the mark for Hōkūpa'a, the fixed star also known as Polaris, and below it was a mark for Newe, the cross also known as the Southern Cross. The bowl also contained three groups of star markings: *nā alanui o nā hōkū ho'okele*, the great paths of the navigation stars used by navigators to guide the canoe across the ocean; *nā hōkū pa'a o ka 'āina*, the fixed stars of the lands that are considered celestial markers for specific island groups, as Hōkūlea is the star over the Islands of Hawai'i; and *nā hōkū i ka lewa a me ka lipo*, the stars of the sky and the darkness, which includes the stars of the southern hemisphere that are not visible in Hawai'i but are necessary to travel to Tahiti (Poepoe 1906).

Although this is only a portion of the training for which a navigator must prepare, it is at least consistent with information Hawai'i anthropologist Sam Low shares in his book, *Hawaiki Rising*,[15] about the training of Mau Piailug, a Micronesian *pwo* from Satawal, sacred master of both the technical and spiritual aspects of Micronesian non-instrument ocean navigation.

Much of the text that follows in this section relies heavily on Low's book and the information shared at the Polynesian Voyaging Society website. The Polynesian Voyaging Society, founded in 1973 by Herb Kāne, Ben Finney, and Tommy Holmes, is a nonprofit research and educational corporation established to research and perpetuate traditional Polynesian voyaging methods.

Mau began learning about the ocean as an infant when he was placed in tide pools to embrace and embody the knowledge of ocean movement (Kawaharada 2010). He was only four or five years old when his grandfather, a famous *pwo*, began teaching him as an apprentice navigator. He started by listening to rhythmic astronomical presentations of several lists of star names by compass quadrant in order of their appearance from east to south, then by reciprocals or groupings, then by sailing directions to islands, and finally by island alignments. After he mastered these teachings, Mau learned the system of *etak*, a Micronesian form of triangulation using the island alignment stars and the swell patterns to dead reckon his position (Low 2013, 54–55). He learned more about star navigation, fishing, and canoe building from his father after his grandfather died but had to complete his training as a *palu*, master navigator, under another *pwo* when his father died. After nearly two decades of learning from other *palu*, *pwo*, and those natural elements and processes associated with the ocean, Mau received the title of *pwo* (Brown 2010).

Two decades later, Mau was approached by Finney to help navigate *Hōkūle'a*, a sixty-foot *wa'a kaulua* built by the Polynesian Voyaging Society to test their theory that purposefully planned long-distance non-instrument ocean navigation was possible. By this time, life on Satawal had changed dramatically. The old ways were being replaced by the ease of modern living. As a result, fewer young men came to learn the art of navigation, and Mau knew the knowledge he held of his craft was near the brink of extinction. So, when Finney offered Mau a chance to be part of the legacy of *Hōkūle'a*, he agreed. He knew this was "an opportunity to pass on his art and by so doing to not only revive Hawaiian culture—but to save his own. *If my people regard the outside world with awe,* he thought, *what would happen if they saw that outsiders valued my knowledge of navigation even more?*" (Low 2013, 191, italics original).

Recruiting Mau, the youngest of six known deep-sea non-instrument navigators at the time, was the only option if the project was to succeed (Low 2013, 50). Mau came to Hawai'i, helped build *Hōkūle'a*, trained with the crew, and, in 1976, navigated *Hōkūleʻa* to Tahiti without the use of modern instrumentation. The crew took thirty days, confronting both westward-moving winds and currents to find an island nearly 500 miles east of their origin, traveling over 2,700 miles. However, fighting the winds and currents was not nearly as bad as the tensions that erupted among the crew, which prompted Mau to leave before the return voyage. This news was extremely devastating for Nainoa Thompson, considered no more than the lowest-ranking seaman at the time, who was looking forward to learning from Mau on the return voyage (Low 2013, 104–105). It was on the voyage home that Thompson created a list of things he wanted to learn, including the changes in the stars, understanding how to use ocean swells as guides, and relating the sounds of the canoe to the direction of the swell (Gordon 2006).

In the months following that return voyage, Thompson invested personal time learning about the stars at Kuliʻouʻou Park on the northeast shores of Oʻahu, staking out a sixty-foot circle with penlights at the cardinal points to create a Hawaiʻi version of Stonehenge in order to observe the night sky (Low 2013, 137). He formed alliances with Will Kyselka, a lecturer at the Bishop Museum planetarium, to further augment his star knowledge and, years later, with the help of Mau, developed the Hawaiian Star Compass, based on the Micronesian star compass (Low 2013, 147, 188–193). It is a diagrammatic representation of a navigator's mental construct of the night sky.

The simplified illustration shown in figure 9 depicts cardinal directions based on the movement of the sun, divides the horizon into quadrants based on types of winds, and delineates each quadrant into seven houses based on environmental navigation cues. The original and complete Hawaiian Star Compass can be found at the Polynesian Voyaging Society website. It includes the locations where stars appear to rise when viewing them in Hawaiʻi at 20 degrees north latitude (the terms used for the cardinal directions are described in chapter 2). Briefly, the term *hikina*, east, designates that part of the horizon where the celestial entities arrive, and the term *komohana*, west, designates that part of the horizon where the celestial entities enter the realm below the horizon. *ʻĀkau*, north, describes that part of the sky where the sun resides during the hot summer season, and *hema*, south, is a term considered the opposite of *ʻākau* and is thus that part of the sky where the sun resides during the cooler winter season.

Thompson divided the horizon into four major quadrants, naming them according to the major wind associated with each quadrant. *Koʻolau* is the

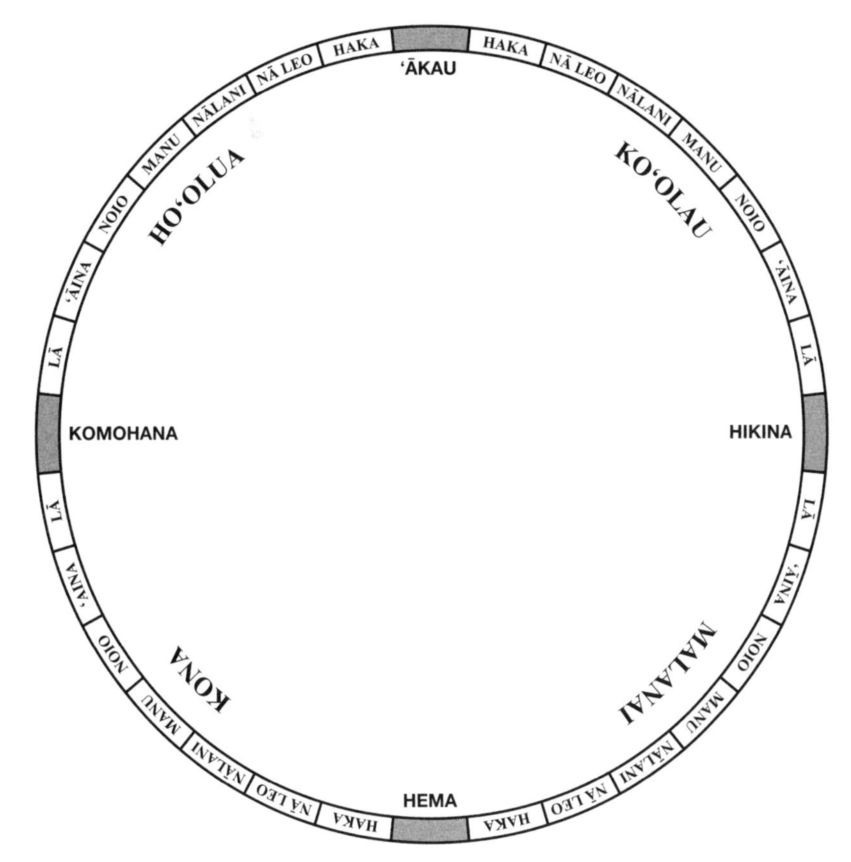

FIGURE 9. Simplified diagram of Thompson's Hawaiian Star Compass (Copyright C. Nainoa Thompson, used with permission)

northeast quadrant and is named for the windward side of the island chain where the trade winds originate. *Malanai* is the southeast quadrant and is named for the gentle breeze associated with the southeast part of Oʻahu and Kauaʻi. *Kona* is the southwest quadrant and is named for the leeward side of the island chain, where it is protected from or in the lee of the constant blowing of the trade winds. *Hoʻolua* is the northwest quadrant and is named for the strong north wind associated with storm systems passing the islands.

Each quadrant was delineated into seven houses to further aid in learning the movement of stars in the celestial sphere. Thompson named them in a manner that reflected the kind of environmental knowledge a navigator must master, such as the winds, birds, and intensity of focus. Starting from the equator and moving northward or southward are *Lā*, *'Āina*, *Noio*, *Manu*, Nālani, *Nā Leo*, and *Haka*. *Lā* is the closest to the equator and represents the region where the sun spends most of its time throughout the year. *'Āina* is the area where land is found, as Hawai'i is at about 20 degrees north and Tahiti is at about 17 degrees south. *Noio*, the Hawai'i noddy tern, is a small black bird that lives on fish and helps a navigator find islands because it flies out to sea in the morning and returns to land at night to rest. *Manu*, bird, is a metaphor for the canoe as it travels on the ocean above the marine life, much like a bird flies in the atmosphere above the terrestrial life. It marks the midway point between the cardinal directions. *Nālani* is the house associated with the second brightest star in the sky, *Ke ali'i o kona i ka lewa*, also known as Canopus. *Nā Leo*, the voices, refers to the voices of the stars speaking to the navigator and acknowledges the intimate communication Kanaka Hawai'i have with nature. *Haka*, empty space, a recipient as a medium or oracle, to stare, is a reference to the emptiness of space near the celestial poles because of the lack of stars in the region (Evanari 1995, 74–75).

The Hawaiian Star Compass gave Thompson a foundation from which he could create a simplified navigational vernacular. It is extremely helpful to know the location where stars appear to rise when you are on an island. However, when you are on a moving vessel with no frame of reference except the location of celestial bodies, being able to measure the movement of specific stars, such as Hōkūpa'a or Newe, in relation to the horizon is integral in determining latitude. To facilitate vertical measurement, Thompson devised a hand measuring stick to determine the angle of stars above the horizon. He calibrated a telephone pole in front of his home using adhesive tape for every degree, with the help of a friend knowledgeable in using a sextant. Then he aligned his hand with the horizon both palm in and palm out and took note of where the various features on his hand lined up with the markings on the pole. He then applied this to the Southern Cross. When it appeared to be at its meridian—where the top star, Gacrux, was in line with the bottom star, Acrux—Thompson measured the distance between the horizon and Acrux as 6 degrees. This would be true for anyone located near 20 degrees north latitude; at 17 degrees latitude, the distance between the horizon and Acrux will be 10 degrees (Low 2013, 140–142).

Thompson had an opportunity to test these methods on an experimental education voyage to determine the rationality of Gordon Pi'ianaia's theory of using the Kealaikahiki Channel as a starting point to travel to Tahiti. Pi'ianaia was the captain on *Hōkūle'a*'s 1976 return voyage from Tahiti to Hawai'i and a veteran seaman. The Kealaikahiki Channel passes between Lāna'i and Kaho'olawe on the leeward side of Hawai'i Island, which provides shelter from the force of the northeast trade winds. This would allow for a calm period of adjustment for the crew, as opposed to the demoralizing beating the 1976 crew endured on its way to Tahiti. The experiment was designed to be a short voyage, traveling on the lee side of the islands until the canoe crossed the path taken in 1976 at about 17 degrees north latitude (Low 2013, 136–140).

The Kealaikahiki voyage began at Mānele Bay on Lāna'i. They made good time using the northeast trade winds until they reached the lee side of Kaho'olawe, where they fell into Maui's wind shadow. The winds picked up again when they reached the 'Alenuihāhā channel between Maui and Hawai'i Island. By this time it was nightfall, and Thompson used the stars to navigate and charted a course to avoid as much of Hawai'i Island's large wind shadow as possible by staying approximately forty miles west of the island. At about three o'clock in the morning, Thompson noticed the wind shifted to the east, a phenomenon known as wraparound winds caused by Hawai'i Island's high peaks. Soon after sunrise the winds shifted back to the northeast, and he knew they had cleared Hawai'i Island's wind shadow. At midnight, he could clearly see the Southern Cross and, using his hand measurement stick, determined they had reached their destination. Upon successfully proving the validity of using the Kealaikahiki Channel, Thompson realized intellectualizing navigation was far different from engaging in it, and he was certain he wanted to learn more than ever (Low 2013, 142–146).

Thompson continued making incremental improvements to his navigational knowledge, but with thousands of stars in the sky he needed to create a simplified means of reading star patterns at different latitudes in both the northern and southern hemispheres. After studying star movements for hundreds of hours both on land and in the Bishop Museum planetarium, Thompson began seeing star pairings at different latitudes, committing them to memory. For example, when the Little Dipper stands upright with its ladle nearly perpendicular to the northern horizon, and Hōkūpa'a is equidistant between the horizon and Kochab, the star that forms the top end of the dipper's bowl, then the observer is at 16 degrees north latitude (Low 2013, 260).

In January 1978, the Polynesian Voyaging Society decided to undertake another voyage to Tahiti, and Thompson knew he needed a streamlined method of memorizing important celestial cues in order to simplify the computations necessary to determine position. He organized the night sky into four star line families:[16] Kekāomakaliʻi, the bailer of Makaliʻi; Kaiwikuamoʻo, the lizard backbone; Mānaiakalani, the name of Mauiʻs fishhook; and Kalupeakawelo, the kite of Kawelo. Each star family occupies roughly a quarter of the celestial sphere, making it possible for the navigator to imagine the celestial sphere when cloudy nights permit limited visibility. Kekāomakaliʻi can be seen in the night sky during the winter season, roughly October/November through April/May. Kaiwikuamoʻo is the next star family and can be seen in the night sky between the winter and summer seasons, roughly January/February through July/August. Mānaiakalani is visible for most of the night during the summer season, roughly April/May through October/November, and Kalupeakawelo is seen in the night sky between the summer and winter seasons, roughly July/August through January/February. Since the departure was set for mid-March, Thompson concentrated on recognizing Kekāomakaliʻi and Kaiwikuamoʻo in the night sky. Even with all these handy self-taught methods, Thompson knew he was barely ready to navigate a new crew back to Tahiti (Polynesian Voyaging Society 2014b).

As it turned out, all his training did not prepare him for the events that occurred on that fateful voyage. *Hōkūleʻa* capsized in the Molokaʻi channel. The crew spent nearly two days in the water and all but one of them, professional surfer and waterman Eddie Aikau, was rescued. Eddie attempted to go for help but was never seen again (Low 2013, 158–175). It was a tragic loss that devastated everyone. While many were consumed with grief and ready to put *Hōkūleʻa* in a museum, Thompson made a courageous decision. He committed himself to becoming the leader necessary to both honor Eddie's spirit to travel the path of the ancestors and shoulder the responsibility of reconnecting with Hawaiʻi voyaging practices. He accomplished this with the help of his father, Pinky Thompson, whose auspicious words at a Polynesian Voyaging Society meeting outlined the organization's aspirations and affected the lives of the people who committed to its dream (Low 2013, 180–182). Thompson continued working with Kyselka but knew in his *naʻau* something was missing. He needed a teacher, someone to mentor him in the things he could not learn on his own. He needed Mau.

Thompson found Mau in Saipan and humbly asked him to help. A few months later Mau agreed and flew to Hawaiʻi to train Thompson, adapting his traditional mentoring style to accommodate the situation. After all, Thompson was a mostly self-taught young man who had used Western technologies and his own ingenuity to learn all he had up to that point. Instead of spending years teaching the stars and their prayers by repetition, Mau adjusted his traditional training methods and allowed Thompson to write down the conceptual framework of the navigator's sky compass before he began a "comprehensive ocean immersion program" (Low 2013, 188–202).

Mau and Thompson spent many days on land and at sea, where Mau taught Thompson how the swells determine direction and predict the approach of an island, how the clouds and color of the sun help foretell weather patterns, and how atmospheric disturbances can signal the location of islands. Thompson could easily identify large waves generated by local winds. However, the smaller swells generated by distant storms or steady winds over time were confusing, which was rather upsetting. He knew an embodied knowledge of ocean movement would mean the difference between life or death, as it could possibly be the only cue a navigator had on cloudy days, the kind of days spent in the dreaded Doldrums or Equatorial Region. So Thompson enlisted the help of David Stroup, University of Hawaiʻi oceanographer, to help him make sense of the patterned directions of different kinds of wave series. He learned to ignore individual waves and look for overall patterns. For example, big wave trains indicate steady winds originating at a distance, whereas scattered waves indicate closer proximity to the wind source (Low 2013, 202–203).

Thompson also learned that classroom teaching is different from the mentoring process Mau used to impart his knowledge. There was no grading or praise involved, but Thompson was eventually able to detect subtle differences in Mau's demeanor that indicated he was progressing in his abilities. For example, they would spend long hours at sea on Thompson's small fishing boat without speaking much at all, but Thompson knew Mau was heavily scrutinizing his ability to navigate. He eventually learned to watch Mau's body language to see if he was relaxed or not and used that as a sign he was indeed making progress. Then one evening Mau asked Thompson to maintain a particular star path and went below to sleep. A curious and overanxious student, Thompson wanted to feel the wave patterns along a different course, only to be greeted by a slightly irritated Mau the next morning. Mau knew Thompson went off-course because of the sound and feeling of the wave swells. In that moment Thompson realized that seeing the swells was not the only way a master navigator acquires knowledge. The multisensual depth of knowledge a *pwo* navigator embodies was something that could be acquired only by spending time on the ocean, experiencing the sights, sounds, and sensations face-to-face (Low 2013, 204–205).

While reading or sensing ocean swells is necessary for increasing the accuracy of position detection, knowing how to read weather patterns provides the crew ample opportunities to prepare for bad weather. Mau taught Thompson how to predict the weather by passing on knowledge he had learned from his grandfather. Together they traveled to the eastern end of O'ahu in the mornings and to the western end in the evenings, hundreds of times. Mau explained what subtle shades of color in the clouds meant: red clouds bring rain, pink clouds indicate fair weather, and dark-blue clouds suggest squalls. Mau also taught Thompson the value of reading the hazy layer near the horizon; heavy haze means heavy winds, an orange tinge indicates lots of wind but no rain. Through this training, Thompson realized that weather prediction was not just about what the sky is showing on one day. It is not a snapshot of conditions at a particular time of day. It seemed as though Mau related to the weather as a living entity, thereby making predictive accuracy better by observing patterns over several days (Low 2013, 206–207).

Mau and Thompson observed the stars, swells, sunrises, and sunsets from land and at sea for the better part of a year. Then one morning, Mau asked Thompson a series of questions about the upcoming voyage, the last one being, "Can you see the island?" Thompson carefully thought about the question, knowing he could not physically see Tahiti, and answered that he could see an image of the island in his mind. Mau told him that as long as he could see the island inside of him, he would never be lost. Without any pomp or circumstance, that conversation effectively ended Thompson's apprenticeship. The rest of his learning would come from the elements of nature themselves. Mau prepared Thompson for those lessons that would deepen his multisensual experience (Low 2013, 206–209).[17]

While there is still so much more involved in the preparation for a long-distance voyage and the training of a navigator and the crew, this section was never meant to be an exhaustive expression of them. It is only meant to provide the reader with an example of the enormity of the task, the time it requires, and the intimacy of the relationships involved in *hoʻokele*. Although each voyage is a saga unto itself, fraught with unknown possibilities, by the time of departure the master navigator has no doubt in his or her ability to see the destination because they carry it within them. The next section is a brief description of Thompson's first voyage to Tahiti in 1980, highlighting several of the cartographic processes of *hoʻokele*.

Ka Huakaʻi (The Journey)

Safely navigating over two thousand miles without the use of modern navigational instruments for the first time is terrifying, no matter how much a person prepares. Thompson had carefully planned each portion of the voyage, accounting for wind and waves, trained with the newly handpicked crew to build bonds of trust with each other and with *Hōkūleʻa*, and committed to memory several mathematical and astronomical shortcuts to assist in position detection. Of these things Thompson was confident; what frightened and exhilarated him were the unknowns he was certain to face in the Doldrums, where heavy cloud cover would obscure the view of the heavens, and the westward-moving winds and waves would certainly push them farther west than they needed to be. Would his preparations be enough to safely navigate *Hōkūleʻa* and crew to Tahiti without irreparable harm?

Adding to his emotional turmoil was the unexpected weather that delayed the launch in Hilo, putting many on edge, as they had schedules to keep or jobs they took time off from or jobs they left to participate in the voyage. Hawaiʻi master woodcarver and crew member Sam Kaʻai recognized the delay as a lesson in humility and an opportunity to recalibrate internal time clocks as the crew learned to move to "the rhythm of the *holo moana*—not the rhythm of the world we live in today" (Low 2013, 239, italics original). Thompson carried these concerns with him, knowing he alone could decide when they could depart, and his decision was based on favorable characteristics of sea and sky. He found comfort in his memorized shortcuts and mental maps, those internal representations of a cognitive cartography.[18] Thompson knew he had to adapt those mental constructs accordingly for the voyage to be successful. This section highlights some of the cognitive cartographic processes he maintained as he navigated to Tahiti for the first time.[19] It is arranged according to the three components of modern-day non-instrument navigation of long-distance ocean voyages: designing and setting a reference course, maintaining the reference course, and locating the destination (Finney 1998, 455; Polynesian Voyaging Society 2014c).

Designing and setting the reference course

Sailing from Hawaiʻi to Tahiti is very challenging because Tahiti lies to the south-southeast of Hawaiʻi, requiring the navigator to travel east against prevailing winds and currents that can divert the canoe about twenty miles westward each day for most of the journey. Thompson spent innumerable hours hunched over his kitchen table studying star, wind, and current patterns late into the night so he could plot his reference course for the voyage. He broke the course up into four segments based on the average prevailing winds and currents. The first segment went from Hilo at about 19 degrees north latitude to about 9 degrees north latitude and made use of the predominant northeast trade winds and west-flowing north equatorial currents. The second segment, from about 9 degrees to 5 degrees north latitude, was the beginning of the Doldrums, where both winds and currents would be unpredictable. The third segment, from 5 degrees north latitude to the equator, was the second half of the Doldrums, where the winds would remain highly variable, but an east-flowing equatorial countercurrent would be less unpredictable. The last segment ran from the equator to Tahiti at about 17 degrees south latitude, passing through the Tuamotu Islands, where the southeast trade winds and west-flowing south equatorial currents would emerge (Low 2013, 213; Thompson 2012).

Thompson plotted each segment accounting for known conditions of wind and current and factoring in the windward capability of *Hōkūleʻa*.[20] For example, on the first segment, strong northeast winds carry the canoe traveling on an *apparent course* of 132 degrees (clockwise from north) about 8 degrees west, making the *course over water* 140 degrees. The *course over water* only takes into account the effect of wind moving the canoe. It does not take into account the fact that the ocean is also moving in a westward direction, carrying the canoe farther west. Based on the average westward-flowing current experienced over the estimated six days necessary to complete the first segment, the canoe would be pushed another 4 degrees west, giving the final *reference course* of 144 degrees. Thompson computed reference courses for all four segments and committed these to memory, knowing it was highly unlikely that the canoe would be able to maintain the reference course throughout the journey, as nature is variable at best. Nonetheless, for his first time as navigator for this long-distance ocean voyage, he knew staying as close to the reference course as possible would make the variables that make up the complex cognitive cartography for the journey far easier to deal with.

Maintaining the reference course

Ideally, the navigator uses celestial entities such as the sun, moon, stars, and planets as the main guide for determining direction. According to Thompson, the sun's first light is the most important time of the day for a navigator to observe the natural elements. Although the exact direction of sunrise varies depending on what time of year it is, the navigator can use it to identify where east is, determine the direction swells originate from, and can feel and hear how those swells interact with the canoe, storing them as multisensual inputs of an embodied memory (Murphy 2015). Furthermore, the sky color, cloud formations, characteristics of the haze near the horizon, and wind directions at sunrise help the navigator determine the weather, which ultimately affects the kind of activity the crew will be undertaking for the day. Less eventful days give the off-duty crew an opportunity to relax, reflect, recharge, and recreate with each other, whereas more eventful days could have all the crew standing at the ready, prepared to take in the sails should a squall loom on the horizon.

In order to stay as close to the reference course as possible, Thompson committed to memory mathematical shortcuts to estimate both total distance traveled and longitudinal, or east–west, course deviation. He calculated distance traveled based on the time a bubble moved across a known distance on the canoe and had already worked out a chart correlating the time in seconds to knots traveled. For example, a bubble taking ten seconds to move the 42.5 feet from bow crossbeams to stern crossbeams meant the canoe was traveling 2.5 knots, eight seconds correlated to 3 knots, six seconds to 4 knots, and so on (Low 2013, 143; Polynesian Voyaging Society 2012a). To deal with longitudinal course deviation, Thompson created an illustration based on the houses of his Hawaiian Star Compass and used an average sailing day of one hundred miles for ease of calculation.

If a canoe set out to travel south but deviated one house east, it would travel only ninety-eight miles in a southerly direction and twenty miles east of the reference course. To compensate for this course deviation, the canoe would have to travel the same distance at a bearing of one house to the west of south. In reality, a canoe travels several different courses in a twelve-hour period, making the computations for course deviation extraordinarily difficult. Adding to the issue of course deviation corrections is that the reference course is based on estimates of average wind and current conditions. If winds allow the canoe to pass the first section in five days instead of six, that means the westward current would have had one less day to push the canoe westward, and it would be farther east than anticipated. The opposite is true if the first section took an extra day to complete. These are the kinds of cognitive cartographic issues Thompson attempted to simplify with his memorized mathematical shortcuts. If he lost track of his longitudinal position in relation to his reference course, he would not be able to reorient

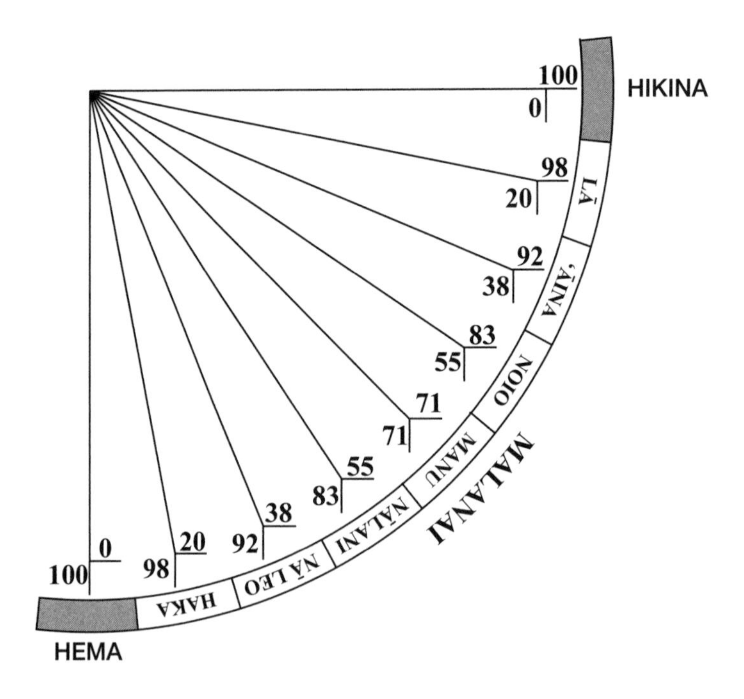

FIGURE 10. Thompson's Course Deviation Math Mantra Diagram based on Low 2013 page 254

himself without some other frame of reference, such as a landmark or seamark (Low 2013, 216, 254; Polynesian Voyaging Society 2012b).

Unfortunately, that is exactly what happened. By day three of the voyage to Tahiti, *Hōkūle'a* had experienced mainly easterly winds that took it eight houses, or one hundred miles, west of Thompson's planned course, and he was concerned that his cognitive cartographic capabilities could not continue to accommodate the ever-increasing adjustments as they drifted farther from his reference course. He knew that when Mau navigated the 1976 voyage and encountered a similar situation, he chose to continue south to the latitude of Tahiti and then tack or sail eastward into the wind in a zigzag pattern to search for the islands upwind. But this was Thompson's first attempt at navigating to Tahiti. Deviating from the planned course would add unnecessary stress to both his mental map and his analytical processes. He pulled Mau aside for a consultation, but Mau recognized that Thompson needed to find his own way to navigate.[21] So, Thompson made the decision to tack back to the reference course, reducing the stress on his mental mapping. Back at the University of Hawai'i, where *Hōkūle'a*'s progress was being transmitted, the resulting map of course deviation and correction very closely matched Thompson's mental map (Low 2013, 254–257).

Holding the reference course requires the navigator to align "the rising or setting sun to marks on the railings of the canoe. There are 8 marks on each side of the canoe, each paired with a single point at the stern of the canoe, where the navigator is stationed, giving 32 bearings to match the 32 directional houses of the Hawaiian star compass" (Polynesian Voyaging Society 2012c). So, if the navigator wanted to head Manu Malanai (SE) or 135 degrees, and the sun rises in the east, then the canoe would be lined up four points forward from the port beam at sunrise.

However, the exact direction the sun appears to rise changes depending on the time of year. It only appears to rise due east on the spring and fall equinoxes, approximately March 21 and September 23, respectively. It appears to rise at 'Āina Ko'olau (ENE) on the summer solstice, approximately June 21, and at 'Āina Malanai (ESE) on the winter solstice, approximately December 22. How would the navigator line up the canoe under these circumstances?[22] What would be the bearing of the canoe if the navigator used the location of the setting sun?[23]

Using celestial entities at night to hold course is a similar process. Since the navigator knows what house the stars appear to rise from, maintaining the reference course means aligning the canoe according to the location of

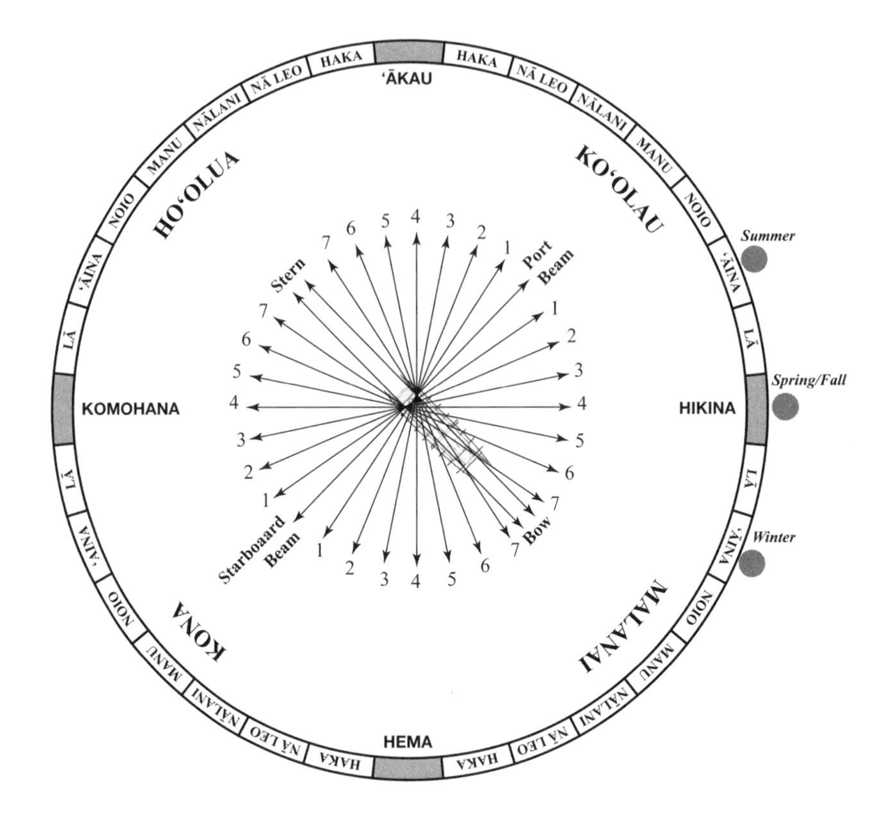

FIGURE 11. Aligning the Canoe to Hold Course (Copyright C. Nainoa Thompson, used with permission)

the house in which the star appears to rise and set. For example, if the navigator could see Hōkūpaʻa, the only star that does not appear to move and remains stationary in the north, and wanted to head Manu Malanai (SE), then the canoe would be lined up four bearings back from the port beam. Using other stars requires the navigator to understand the stars' curving paths through the celestial dome, which consists of three components: the house a star appears to rise from and set in, the latitude of the canoe, and the star's rate of movement in the sky.

As described in chapter 2, all celestial entities appear to ascend and descend at an angle, depending on where you are viewing them from. For example, in Hawaiʻi, stars appear to rise at an angle of 20 degrees south of a line perpendicular to the equator. However, in Tahiti, the same stars appear to rise at an angle of 17 degrees north of a line perpendicular to the equator.

The navigator must also take into account that celestial entities appear to rise and set at a rate of about 15 degrees in an hour. So, after two hours it becomes difficult to determine where a star appeared to rise. This is why the best time to use stars to determine position is when they are near the horizon (Polynesian Voyaging Society 2012c, 2014a, 2014b).

Navigators can also use the moon to determine position, but they need to keep a few things in mind. The moon rises about forty-eight minutes later each night and appears to rise from a different place on the horizon, between 'Āina Ko'olau (ENE) and 'Āina Malanai (ESE). As the moon

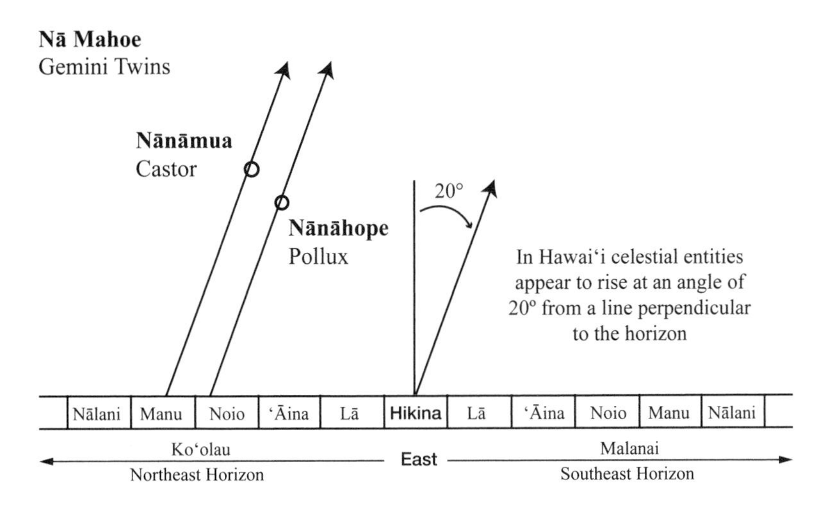

FIGURE 12. Nā Mahoe star family from Hawai'i

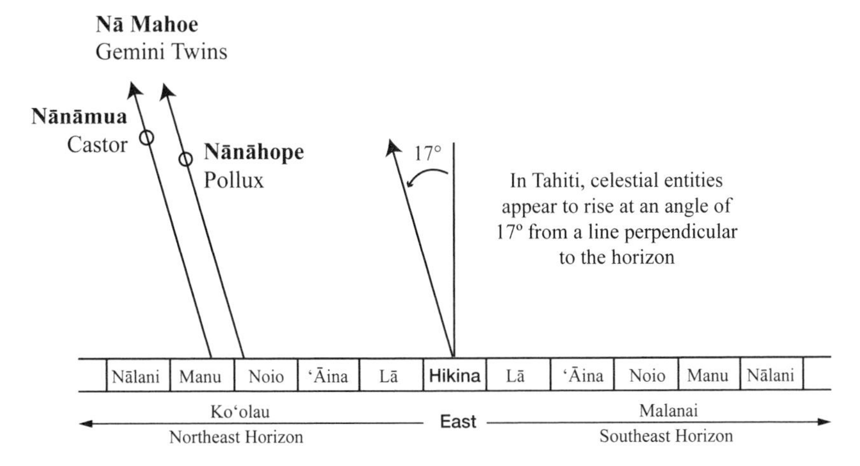

FIGURE 13. Nā Mahoe star family from Tahiti

waxes and wanes, the line separating the dark side of the moon from the lighted side of the moon approximates a north–south alignment. Lastly, the moon does not circle the earth on the same plane as the sun. So, as the canoe moves from the northern to the southern hemispheres, the location on the horizon from where the moon appears to rise changes (Polynesian Voyaging Society 2012c).

When the sky obscures celestial clues, the navigator must use ocean swells to determine direction. This is why it is essential to note the direction of the rising and setting sun, as it is the largest, brightest celestial entity. Ocean swells are different from ocean waves. Swells are waves that have traveled long distances and tend to move in straight lines from their origins. They are also more stable than waves that are generated more locally, which tend to be choppier. Some navigators are able to feel the swells better than see them, as their effect on the canoe produces rhythmic movements. Orienting the canoe according to the pitch (front-to-back movement) or roll (side-to-side movement) means matching the movement of the swell to the alignment on the canoe, as described above. For example, maintaining a heading of Manu Malanai (SE) with an ocean swell coming from the north means aligning the movement four points back from the port beam. Complications arise in the Doldrums, where an abundance of fast-moving squalls generate wave patterns that are difficult to determine. Add to this complication the unfortunate situation of not being able to see the sky for a few days, and you have the exact scenario Thompson faced on day eleven, when *Hōkūleʻa* entered the Doldrums.

The sky and ocean had gone black. Strong winds and constantly shifting waves made the canoe move quickly in an unknown direction. By this time Thompson was mentally exhausted, physically fatigued, and barely holding himself together; he had no idea where to direct the crew. In a moment of terrified calm, he accepted that answers could not come from his intellect and he let go of his fears of failure. In that moment, he connected with an instinctual depth of sensual knowledge beyond the analytical or physical and just knew where the moon was located. Something inside him felt its pull over his right shoulder (Low 2013, 276–277, 324). As Thompson describes the experience,

> I can't explain it. . . . There was a connection between something in my abilities and senses that went beyond the analytical, beyond seeing with my eyes. It was something very deep inside. Before that happened, I relied on math and

> science because it was so much easier to understand things that way. I didn't know how to trust my instincts. My instincts were not trained enough to be trusted. That night, I learned there are levels of navigation that are realms of the spirit. Hawaiians call it na'au—knowing through your instincts, your feelings, rather than your mind or your intellect. It's like new doors of knowledge open and you learn something new. But before the doors open you don't even know that such knowledge exists. (Low 2013, 277)

That night Thompson's sensual awakening opened a door he did not even know existed, because he grew up in a world where he was not taught to trust it. As he navigated to Tahiti for the first time, he knew he could rely on his preparation. The science, math, and planetarium gave him the confidence to take on the challenge, but it was his training with Mau that prepared him to accept new ways of knowing. From that point on, Thompson shed his fear of the Doldrums and began seeing it as one of the most beautiful places on earth; getting through it became less of a burden (Low 2013, 280–281).

What began as tuning in with the rhythmic movements of *Hōkūle'a* became an intimate and trusting relationship between navigator, canoe, and natural entities. Thompson began trusting his instincts and listened to his *na'au* when unexpected currents yet again sent him farther west than his planned course. His planned course gave him a reasonable margin of error. However, as Thompson nurtured this intimate connection with *Hōkūle'a* and the natural entities, his confidence as a navigator grew, and this time he embraced the unexpected course deviation, sensing *Hōkūle'a* communicating *let me go, I know where land is.* He listened and decided to head for Mataiva along the western edge of the Tuamotu Islands, knowing if he missed it, he could completely miss the Tuamotu Islands and could possibly end up sailing into open ocean, further complicating the voyage (Low 2013, 311–314).

Locating the destination

Navigators increase their ability to find their destination by noting the directional alignment of island clusters, creating a target box, and determining the most effective approach. In the Pacific there are several island clusters. For example, the major Islands of Hawai'i stretch from the northwest to the southeast. Approaching them from the southeast creates a narrow target area, but approaching from the east creates a larger target area. On the fourth and last leg of the journey, Thompson's target box for his destination ranged from Manihi in the Tuamotu Islands to Maupiti in the Tahitian Islands, giving him a target area of about five hundred miles in both the north–south and east–west directions. He planned to approach this target from the east, heading for Takapoto in the Tuamotu Islands. Since the Tuamotu Islands also run from northwest to southeast, the strategy increases the probability of successfully locating his ultimate destination, Tahiti (Polynesian Voyaging Society 2012d).

As the canoe approaches the vicinity of any island cluster, the navigator and crew begin looking for signs of land, such vegetative debris, cloud clusters, and birds. Of these signs, diurnal seabirds such as the *noio*, which go out to sea in the morning in search of fish and return at night to rest, are the most useful. However, Thompson learned things change when birds are nesting and taking care of their young. On day twenty-eight of the voyage, the crew spotted a leaf and the kind of seaweed that grows on coral reefs. Thompson estimated they were about two hundred miles east of Mataiva. By day thirty he estimated they were only eighty miles away, but the current was confusing. He expected swells from the southeast, but they were experiencing swells from the northeast, east, and southwest. Then Mau spotted a *noio* flying from south to north in the morning, heading to its feeding grounds. In the evening they saw birds flying south-southeast, but they still could not see land. They waited throughout the night, and a little after midnight, *Hōkūleʻa* slowed to a stop (Low 2013, 315–319).

The next morning, they sighted a *noio* flying south, not north to its feeding grounds. Everything in his *naʻau* told Thompson he was in the right place looking for the right signs of landfall, but nothing made sense, as first the current was not what he expected and now a bird was flying in the opposite direction. In a moment of panic, Thompson thought they had drifted past Mataiva, and he ordered the crew to turn *Hōkūleʻa* around and head north, the direction the *noio* came from. Then Mau pulled him aside and told him to turn the canoe back around and follow the bird. An hour later, they made landfall on Mataiva. After the celebrations calmed, Mau told Thompson the reason the bird was flying back to the island was because it had a fish in its beak and was returning to feed its babies (Low 2013, 319–322).

Perceptions of Kanaka Hawaiʻi Spatial/Temporal Knowledge Acquisition

The voyage ended shortly thereafter, when Thompson guided *Hōkūleʻa* and the crew to Tahiti, a mere two hundred miles away, which barely compared to the over two thousand miles they had already traveled. Guiding a vessel across the single largest natural feature on the earth, the Pacific Ocean, without the use of modern instrumentation, is an analytically, physically, and culturally conscious feat, proven without a doubt to still be possible to this day. The cognitive cartographic processes of *hoʻokele* depend on adapting stored spatial/temporal knowledge to environmental changes acquired empirically through culturally attuned sensory systems.

Through the works of Meyer and Oliveira we learned that Kanaka Hawaiʻi sensory inputs are both culturally distinct and genealogically embodied. Even with all the scientific and mathematical preparations, in Thompson's moment of terrified calm, when he moved past the depths of his own darkest fears, he attributed his moment of clarity to his *naʻau*, that spiritual umbilical cord that serves as a generational link to ancestral knowledge. That shift of his consciousness and certainty of its origin awakened an alternative sensual input that lay dormant, awaiting opportunity to be of service once again. Acknowledging this link to ancestral knowledge was one of the most important contributions Thompson made to Kanaka Hawaiʻi cognitive cartographic processes and spatial/temporal cognition.

Kanaka Hawaiʻi spatial/temporal cognition is based on Kanaka Hawaiʻi philosophical underpinnings developed over hundreds of years through intimate interaction with the natural environment. For example, being surrounded by the elements of nature, with the various layers of sky above and the ocean below, while floating on a *waʻa kaulua* imbued with *mana* through intentional prayer according to traditions tried and true, and led by a navigator trained to process external stimuli as though they are road signs marking the directions to their destinations, made the 1980 voyage more than just another step in proving a theory of Polynesian navigational competence. Aside from the boost it gave the Hawaiʻi cultural resurgence decades ago, the 1980 voyage proved that a Kanaka Hawaiʻi cognitive cartography is capable of acquiring sensory input from a collective consciousness embedded in our genetic makeup.

As descendants of navigational peoples, Kanaka Hawaiʻi inherited nearly three millennia of celestial knowledge. This knowledge is being reinvigorated by the determination and dedication of people and organizations like the Polynesian Voyaging Society. Their story is a modern-day epic, with heroes and heroines guided by ancestral nudging and a "knowing" in their *naʻau* that long-distance non-instrument ocean voyaging was dangerously close to being lost. Like so many epics, great successes required great sacrifice, and this one had many. While we may never know with great certainty what it was like to navigate across the Pacific the way my ancestors did it hundreds of years ago, it does not mean that what is being accomplished today is any less a Hawaiʻi practice. The tools may have changed because technology changed, and adapting to those changes is necessary for cultural survival. Thompson didn't have someone to teach him star groupings and constellation locations using an *ipu hōkeo*. So he learned them with the aid of a planetarium, and nowadays students have access to astronomy programs and apps. Using these new tools and technologies to teach, learn, or implement *hoʻokele* is still very much a Hawaiʻi practice, because it still reflects a Kanaka Hawaiʻi spatial/temporal cognitive consciousness.

Hoʻokele may not be an exacting science, but a successful voyage is not measured by precision or accuracy. It is measured by safely arriving at the destination without irreparable harm, by the bonds created between beings both human and other-than-human, by the respect for rhythmically tuning in with something other than schedules and agendas, and by the depth of understanding accessed within each person, whether of self or of the essence of the practice. *Hoʻokele* continues to evolve as a blend of Hawaiʻi-based techniques and Western technologies. The Polynesian Voyaging Society now offers links to programs offering various forms of Hawaiʻi navigational training, and there are several websites offering exercises that can be incorporated into school curricula.

Although the next generation of potential Hawaiʻi navigators continues to benefit from Thompson's personal journey, he explains that becoming a navigator is not for everyone, as it is very specialized and a deeply personal journey (ʻŌiwi TV 2014). It requires a person intellectually capable in the classroom, on the land, and at sea; a person willing to commit a lifetime extending that intellectual learning into embodied learning by building personal relationships with the ocean, the atmosphere, and the celestial dome; and a person courageous enough to step beyond the known limits of Western scientific paradigms and embrace the spirit of the voyage, where their entire being inextricably intertwines with a divine consciousness. Such a person recognizes their responsibility to the knowledge they learn and reciprocates, giving it back to their community and the next generation, continuing the cycles of learning and teaching.

As I write these words (2015), *Hōkūleʻa* is leaving Auckland, Aotearoa, for Sydney, Australia, as part of the Polynesian Voyaging Society's Mālama Honua Worldwide Voyage, carrying the next generation of *hoʻokele* practitioners in training as well as the very essence of the Hawaiʻi culture, much like Pele did when she traveled from Tahiti to Hawaiʻi as expressed in the opening chant. *Honuaiākea* and *Kānekālaihonua*, Pele's canoes, did not carry only her family; they also carried her knowledge traditions and understandings of the world as well. *Hōkūleʻa* carries with it Hawaiʻi knowledge traditions and understandings as it promotes the message that the world must work together, bridging traditional practices with modern technologies to become better stewards of our "island earth."

CHAPTER FIVE

Ka Haku ʻAna

(Hawaiʻi Composing)

"Hilo Hanakahi"
Composed by Keola Nālimu

Hilo Hanakahi
Hilo of Chief Hanakahi
I ka ua Kanilehua
Rustling with the Kanilehua rain
Puna paia ʻala
Puna of the fragrant forest clearing
I ka paia ʻala i ka hala
Fragrant with the blossoms of hala
Kaʻū i ka makani
Kaʻū of the wind
I ka makani Kuehulepo
The dust stirring wind of Kuehulepo
Kona i ke kai
Kona of the sea
I ke kai māʻokiʻoki
The sea with the streaking patchwork of hues
Kawaihae i ke kai
Kawaihae of the sea
I ke kai hāwanawana
The softly whispering sea
Waimea i ka ua
Waimea of the rain
I ka ua Kipuʻupuʻu
The pelting Kīpuʻupuʻu rain
Kohala i ka makani
Kohala of the wind

I ka makani ʻĀpaʻapaʻa
The strong buffeting ʻĀpaʻapaʻa wind
Hāmākua i ka pali
Hāmākua of the cliffs
I ka pali lele koaʻe
The cliffs where tropic birds soar
Haʻina ka puana
Tell the story in the refrain
I ka ua Kanilehua
The story of the Kanilehua rain
(Wilcox et al. 2003, 65)

Ka haku ʻana, composing, inventing, arranging, braiding, plaiting, or putting words in order, incorporates sensual Kanaka Hawaiʻi cartographic processes reflecting the intimate engagement Kanaka Hawaiʻi have with nature. *Haku* is a term generally recognized as a form of making *lei* but is also metaphorically understood as composing Hawaiʻi verbal arts such as *mele,* Hawaiʻi song, chant, poem; *ʻōlelo noʻeau,* Hawaiʻi proverb; *ʻōlelo nane,* Hawaiʻi riddle, parable, or allegory; *moʻolelo,* Hawaiʻi historical accounts; and *kākāʻōlelo,* Hawaiʻi oratory. A *mea haku* is a general term for a person that does the composing, a person who encodes symbolic meaning through the poetic plaiting of Kanaka Hawaiʻi consciousness. More specific terms include *haku mele*, Hawaiʻi composer of song, chant, or poetry, and *haku moʻolelo*, Hawaiʻi composer of historical records.

Hawaiʻi literary scholar Kuʻualoha Hoʻomanawanui describes *haku* as having metaphorically overlapping meanings as "both *lei* and *mele* are composed, with the *mea haku* (one who composes) selecting, arranging, and putting in order the *pua*—literally, the flowers; in poetry, the metaphors and symbolic imagery that evoke *kaona*" (2005, 32–33; italics added). She also asserts that Hawaiʻi verbal arts reached a highly sophisticated state prior to Western contact, as they were nurtured by a language tradition that valued complex symbolic elements and utilized *meiwi,* literary devices, to aid the composer, performer, and observer/participant. Take, for example, the opening *mele* "Hilo Hanakahi," composed by Keola Nālimu in the 1920s.

It is a *mele aloha ʻāina*, song that expresses deep affection for the land, describing outstanding characteristics of various land areas on the Island of Hawaiʻi. Hilo Hanakahi is an area on the Keaukaha side of the Wailoa River that honors the beloved *aliʻi* Hanakahi, renowned for peace and prosperity during his reign. It is also associated with the Kanilehua rain, a mistlike rain that quenches the thirst of *lehua* blossoms, the flower of the *ʻōhiʻa* tree that grows in the Hawaiʻi forests. Puna is the largest wet forest on the island and is well known for the heady fragrance of *hala*, pandanus tree, blossoms of which were often woven into the walls of *hale*, Hawaiʻi house. Kaʻū is a dry land associated with the dust-stirring Kuehulepo winds. Kona is famous for its nearshore ocean depths creating streakingly colorful variegated seas. Kawaihae sits at the juncture of two shield volcanoes, where a naturally safe and calm bay is known for its softly whispering seas. Waimea occupies the saddle of the same juncture, through which the northeast trade winds are funneled, generating thick, heavy, drenching cascades of the pelting Kīpuʻupuʻu rain. Kohala is at the northernmost tip of the island, where the northeastern trade winds whip around the headland and become the strong, thrashing ʻĀpaʻapaʻa winds. Lastly, Hāmākua is on the windward side, catching plenty of rainfall that accounts for its steep valleys and steep cliff coastline where tropical birds soar.

This *mele* appears to be a straightforward presentation of each area's physiological features and as such is a Kanaka Hawaiʻi cartographic composition sensually orienting a traveler to the dominant features of each area using *wahi pana*, celebrated place. Although some *mele aloha ʻāina* are composed with embedded *kaona* and are performed with the added purpose of being a precursor to a sexual encounter, Pukui cautions that not all *mele* have an added metaphoric layer "and to read such meaning into them is folly" (1949, 251). Nonetheless, the prevalence of quenching and drenching rains, fragrances, winds, and whispering in this *mele* easily allows one's imagination to focus on a more intimate exchange that needs no explanation to reveal the underlying embedded metaphors (I have included my own understanding in the footnotes).[24]

Kanaka Hawaiʻi love composing *mele*, and anyone, regardless of gender, age, profession, or social standing, could and would compose poetic expressions of love, longing, or laments of loss; however, this does not mean that everyone was trained or mentored in *haku*.[25] Like other Kanaka Hawaiʻi cartographic performances, certain types of *ka haku ʻana* require years of preparation. A skilled *mea haku* arranges ideas, embedding symbolic meaning and inspiring the observer/participant to remain sensually engaged with nature. As such, many *mele* maintain sociocultural historical records as well as Kanaka Hawaiʻi scientific perspectives. The next section presents what it takes to become a skilled *mea haku* before describing different types of Hawaiʻi *meiwi,* verbal arts, and symbolic elements embedded in their *haʻiʻōlelo,* oral presentation, that stimulate sensual attentiveness and deepen a sensual awareness of an awakened connection with nature. Much of the information in this chapter is derived from both archival documentation and my personal experiences.

Preparing the Composer

Composing either prose or verse for formal political or informal social events is a highly valued skill among Kanaka Hawaiʻi. It requires the *mea haku* to understand the structure of various Hawaiʻi verbal art forms, incorporate an assortment of *meiwi,* and embed symbolic meaning. Certain types of Hawaiʻi verbal art forms require years of learning to compose, perform, and share compositions capable of spanning generations. Oftentimes in the past children were selected and groomed for these specific types of Hawaiʻi verbal arts at an early age, so they could be taught the old *mele* and the technique for fashioning new *mele* (Pukui 1949, 247).

Like the training for the navigator, training for an orator consisted of listening to and repeating scores of *mele* and *moʻolelo.* According to the Hawaiʻi language scholar Kalena Silva, quoted in an article by Fullard-Leo, a "popular training competition involved two youngsters lying chest down facing the sun beside a placid pool of water. Each inhaled, then slowly whispered, 'naʻu-u-u-u,' while a third judged who could sustain the hum the longest by watching the rippling water" (Fullard-Leo 1997). Controlling the pitch and duration of a breath is very important. Sometimes training sessions lasted hours, and students were asked to imitate the sounds of various natural elements like the waves breaking onshore or the rushing water of a waterfall. These types of competitive training methods were especially important for those orators specializing in or performing ritualistic compositions, which usually contained prolonged phrases that were expected to be completed in a single breath.

Hawaiʻi verbal arts genres

There are several genres of Hawaiʻi verbal arts. Samuel Elbert, Hawaiʻi ethnographer and lexicographer who worked with Pukui, classified them according to those types not meant to be repeated verbatim, such as *kākā'ōlelo, mo'olelo,* and *ka'ao,* a legend, tale, or fanciful fiction, versus those meant to be repeated verbatim, such as *'ōlelo no'eau, 'ōlelo nane,* and all types of *mele.* Scholar of Hawai'i language and anthropology Puakea Nogelmeier identifies different types of *mele* based on content, presentation styles, and vocal techniques.

> Terminology is unevenly recorded and overlapping, but certain genres are generally acknowledged: ko'i honua, mo'okū'auhau, ha'i kūpuna, kāmākua (genealogy and origin chants); mele inoa, mele ma'i, mele ho'āla (name and personal chants); mele pana, mele aloha 'āina (place or loyalty chants); mele pai ali'i, mele ho'ohanohano (chiefly or honorific chants); mele aloha, mele ho'oipoipo (love chants); pule, kau (prayer or eulogy chants); mele nema, paha (criticizing or challenging chants); mele kāhea, mele komo, mele ka'i, mele ho'i (entry and procession chants); kanikau, mele kūmākena, kūō, uē helu (mourning chants). (2001, 3)

Reading through Nogelmeier's examples brings to mind an important point of clarification—the difference between *mele* and *oli.* According to Kanahele, "*mele* refers to sung poetry, and *oli* to the voice techniques used to deliver *mele*" (2001, v). *Oli* is also a specific type of oral performance not meant for dancing. Those *mele* meant to accompany *hula* were distinguished as *mele hula.*

It was the practitioners of *kākā'ōlelo* and *oli* that carefully selected and mentored protégés, as their performances were tied to political and ceremonial arenas, respectively. While much has been recorded and written about *oli,* very little remains available about the art of *kākā'ōlelo.* Hawai'i language scholar Hiapo Perreira dedicated over a decade of his life to the redevelopment of the art of *kākā'ōlelo.* He asserts the rhetorical flourish of *kākā'ōlelo* is more closely related to other Polynesian cultures than to the format seen in typical mainstream American "speech giving" presentations in its "style (verbal and nonverbal), content, spiritual intent, and overall élan!" (2011, v).

Hawai'i researcher and scholar Reverend Malcolm Naea Chun shares that oratory is a highly respected and vital part of many Pacific Island cultures and closely aligns *kākā'ōlelo* with Sāmoan, Māori, and Tahitian traditions. His research indicates the *kālaimoku,* chief counselor to the *ali'i,* was a *kākā'ōlelo* and "had been schooled in not only the history and deeds of former and present chiefs but also in the words they used" (2011, 233). Perreira expands this research considerably, indicating that the *kākā'ōlelo* was "a polymath responsible for voicing historical authority and expertise in all matters of the royal court, ranging from politics to genealogy and history to education, probably also one who jested with humorous wit to entertain in times of leisure and repose" (2011, viii–ix). Unfortunately there is no example of this kind of oratory in either audio recordings or written descriptions in the huge compendium of Hawai'i archival documents. Nonetheless, Perreira persisted and pieced together this form of Hawai'i oratory by examining other Polynesian oratorical traditions and paying careful attention to classical Hawai'i writing styles found in the Hawai'i language newspapers, among other sources.

The outline he provides for creating a modern version of a traditional *kākā'ōlelo* carries certain characteristics similar to those found in Sāmoan and Māori oratory, such as their dramatic style of presentation, low tolerance for error, stringent methods of guiding neophytes into becoming orators, and recognition of the practice as a means to maintain historical accounts. Using these characteristics as guides, Perreira developed a method for reclaiming *kākā'ōlelo* involving contextually appropriate usage of Hawai'i vernacular phrasing and *meiwi.*

Meiwi (literary devices)

Kanaka Hawai'i greatly admired persons with extraordinary memory and performance skills, as it was an important means of maintaining the continuity of past traditions and ensuring future survival. Unlike a society that relies on information stored in books, documents, or archives, whether printed or digital, a society that transmits information orally or performatively places great value on listening and memorization skills. Noelani Arista notes that *mea haku* acolytes "were trained to have retentive and relational memories and were entrusted with portions of a vast oral tradition" (2009). To facilitate memory recollection that also provided pleasing performances, *mea haku* incorporated *meiwi* into their compositions, such as mnemonic structure, repetition, and rhyming schemes. Today *meiwi* remain a distinctive element of Hawai'i *mele.*

Mnemonic Structure

One method *mea haku* use to remember their compositions is to create familiar or repeating formulaic mnemonic structures. With over two thousand lines of text, the Kumulipo is rife with mnemonic devices, providing the orator of the *mele* ample opportunity to succeed in performing it.

Hoʻomanawanui identified several mnemonic devices, including the following "fill in the blank" pattern: *Hānau ka _____, hānau ka [or ke] _____ i ke kai lā holo*, from *wā ʻelua,* the second space/time interval.

"Kumulipo Wā ʻElua"

Hānau ka iʻa, hānau ka naiʻa
Born is the fish, born the porpoise
i ke kai lā holo
in the sea, swimming
Hānau ka manō, hānau ka moano
Born is the shark, born the goatfish
i ke kai lā holo
in the sea, swimming
Hānau ka mau, hānau ka maumau
Born is the mau, born the maumau
i ke kai lā holo
in the sea, swimming
Hānau ka nana, hānau ka mana
Born is the nana, born the mana
i ke kai lā holo
in the sea, swimming
Hānau ka nake, hānau ka make
Born is the nake, born the make
i ke kai lā holo
in the sea, swimming
Hānau ka napa, hānau ka nala
Born is the napa, born the nala
i ke kai lā holo
in the sea, swimming
Hānau ka pala, hānau ka kala
Born is the pala, born the kala
i ke kai lā holo
in the sea, swimming
Hānau ka paka, hānau ka pāpā
Born is the paka, born the pāpā
i ke kai lā holo
in the sea, swimming
(Beckwith and Luomala 1981, 191; explanation from Hoʻomanawanui 2005, 37)

Hoʻomanawanui explains the italicized pairings in the example shown above as rhythmically pleasing "musical sequences of sound . . . [that minimize] the daunting challenge of memorizing and verbally recalling over two thousand lines of text" (2005, 37–38). She finds thirty instances of this particular mnemonic pattern in *wā ʻelua* and a number of other formulaic patterns throughout the Kumulipo.

A similar formulaic mnemonic structure is used in the following *mele*, found in Kaʻao no Laukiamanuikahiki, legend of Laukiamanuikahiki. Laukiamanuikahiki is a beautiful young woman who disguised her looks so she could visit her brother, who also happened to be her husband in their homeland, and her sister-in-law. After being mistreated by them, she reclaimed her beauty and performed this *mele*. While incestuous relationships were not necessarily prevalent, they were also not considered aberrant or immoral behavior in Hawaiʻi society. It is, however, considered rude behavior to mistreat a visitor whom you welcome into your home.

"E Kini o ke Akua"

E kini o ke akua
Ye forty thousand gods
E ka lehu o ke akua
Ye four hundred thousand gods
E ka lālani o ke akua
Ye rows of gods
E ka pūkuʻi akua
Ye collection of gods
E ka mano o ke akua
Ye four thousand gods
E ke kaikuaʻana o ke akua
Ye older siblings of the gods
E ke akua mūkī
Ye gods that smack your lips
E ke akua hāwanawana
Ye gods that whisper
E ke akua kiaʻi i ka pō
Ye gods that watch by night
E ke akua ʻalaʻalawa o ke aumoe
Ye gods that show your gleaming eyes by night

E iho, e ala, e oni, e ʻeu
Descend, awake, appear, animate
Eia ka mea ʻai ia ʻoukou la, he hale
Here is your food, a house
(Fornander 1985, 4: 606–607)

In this *mele* the first six lines have the following pattern: *E (ka/ke)* _____ *(o ke) akua,* where the blanks are filled with different terms describing the numerous, or multitudes of, divine entities. The next four lines use the pattern *E ke akua* _____, where the blanks are filled in with a description of specific types of divine entities. The *mele* ends with a commanding request to come hither, "descend, awake, appear, animate," as an offering of food in the form of a *hale*, Hawaiʻi house, is theirs for the taking. This *kaʻao* ends with the *hale* and her sister-in-law being consumed by fire. While legends such as these hold the promise of exacting extreme forms of punishment that all too often seem to be commonplace these days, the message here about the moral injustice is all too clear: It is never a good idea to mistreat anyone, regardless of his or her station in life.

Repetition

There is a great emphasis on repetition in *mele*, both as *māhuahua*,[26] reduplication or a repetition of syllables, and as *pīnaʻi*, repeated term. Many terms in the Hawaiʻi language are *māhuahua*, which, though it usually implies increased frequency, plurality, or a repeated activity, is not always the case, as it can also represent a diminutive explanation, such as *weliweli*, small sea cucumber, versus *weli*, sea cucumber. Also, certain terms exist only as reduplications, such as *ahiahi*, evening, and *ikaika*, strong (Elbert and Mahoe 1970, 12). There are *māhuahua* in the opening *mele*, "Hilo Hanakahi," at the start of this chapter.[27]

Pīnaʻi, repetitive rhythmic markers usually set at the end of lines, are also rhythmically pleasing musical sequences of sound. In the following *mele*, the *mea haku* uses the *pīnaʻi* to promote a sexual whimsicality.

"Tewe Tewe"

ʻOʻopu nui, tewetewe
Big fish, move back and forth
Taʻa mai ana, tewetewe
Move to satisfy, move back and forth

Pā i ka lani, tewetewe
Touch the sky, move back and forth
Tōheoheo, tewetewe
Tumble down, move back and forth

Hui
(chorus)
Teketeke tewe tewe tewe
Prepare, move back and forth

ʻOʻopu nui, tewetewe
Big fish, move back and forth
Paʻa i ka lima, tewetewe
Caught in the hand, move back and forth
Ke ʻoni nei, tewetewe
Reaching here, move back and forth
Kūpaka nei, tewetewe
Twisting about, move back and forth

ʻOʻopu nui, tewetewe
Big fish, move back and forth
Te tomo nei, tewetewe
Enter, move back and forth
I ta ʻupena, tewetewe
Captured in the net, move back and forth
A kāua, tewetewe
That is ours, move back and forth

ʻOʻopu nui, tewetewe
Big fish, move back and forth
E akahele ʻoe, tewetewe
Take it easy, move back and forth
O hemo aʻe nei, tewetewe
Or you'll get loose, move back and forth
Paʻa ʻole iā tāua, tewetewe
Before we finish, move back and forth
(Wilcox et al. 2003, 252)

The *ʻoʻopu*, goby fish, lives in brackish water and is considered a delicacy, as its sweet succulence is a reflection of the water quality from which the fish was caught. It is often used to represent partners engaged in sexual activity. Here the *mea haku* does nothing to disguise the sexual innuendo, yet playfully provides just enough room for the imagination to run wild.

Rhyming Schemes

Several *meiwi* found in many *mele* are meant to provide an ordered remembering process, including *helu*, to list; *inoa*, name; *kuʻina*, connective joining; and *alolua*,[28] opposites. Both *helu* and *inoa* are found in the *mele* "Hilo Hanakahi," as it names land areas and their accompanying terrestrial or atmospheric entities in a consecutive list. *Kuʻina* are elements that pair sounds from the end of one line with those at the start of the next line, as seen with the underlined words in the first two verses of the *mele* "Hilo Hula," composed by Joseph Kalima.

"Hilo Hula"

Kaulana mai nei ʻo Hilo ʻeā
Renowned is Hilo
Ka ua Kanilehua ʻeā
For the rain that sings upon the lehua
Ka ua hoʻopulu ʻili ʻeā
The rain that soaks the skin
Ka ʻili o ka malihini ʻeā
The skin of the visitors

Nani wale hoʻi ka ʻikena ʻeā
So very beautiful is the view
Ka nani o Waiākea ʻeā
Of the loveliness of Waiākea
Ka wai o Waiolama ʻeā
The waters of Waiolama
Mālamalama Hawaiʻi ʻeā
All of Hawaiʻi is brightened
(Wilcox et al. 2003, 66)

In the first verse we see the term at the end of the third line, *ʻili*, skin, being reinforced by use of repetition at the beginning of the fourth line. In the second verse the terms *wai*, freshwater, and *lama*, an endemic ebony hardwood tree and a torch or light, are reiterated in a similar manner. Aside from being a helpful mnemonic technique, this also creates a pleasing repetitive sound when performed.

The *alolua* is a method of rhyming opposites, as found with the underlined words in the beginning of the *mele* "Nā ʻAumākua."

"Nā ʻAumākua"[29]

Nā ʻaumākua
Ancestors
mai ka lā hiki a ka lā kau
from the rising sun to the setting sun
Mai ka hoʻokuʻi a ka halawai
From the zenith to the horizon
Nā ʻaumakua
Ancestors
ia kahina kua
who stand at our back
ia kahina alo
at our front
ia kaʻa ʻakau i ka lani
at our right side in the sky
ʻO kihā i ka lani
A breathing in the sky
ʻOwe i ka lani
A murmuring in the sky
Nunulu i ka lani
A reverberating in the sky
Kāholo i ka lani
A fast moving in the sky
Eia ka pulapula a ʻoukou ʻo [Mahoe]
Here is your descendant, o [Mahoe]
E mālama ʻoukou [iā ia]
Safeguard [him]
(Malo 1971, 11; explanation by Louis)

A modified version of the *mele* is often used today when a person or group of people request the assistance of divine entities for protection or healing. The bracketed text in the last two lines can be changed to fit with the intended function or purpose. The *mea haku* made sure all divine entities, from rising to setting sun, from zenith to horizon, from the backs to the fronts of our bodies, are included in the request for protection.

These are but a few of the *meiwi* that a *mea haku* uses to create compositions for ease of memory retention. While it appears as though *meiwi* add poetic flourish, they are, in fact, necessary memory aids, especially for longer compositions such as the two-thousand-line Kumulipo, meant for ceremonial or other official public presentation. Inaccurately reciting *mele* in these kinds of venues met with serious life-threatening consequences such as loss of status or loss of life (Tartar 1982, 23). Of equal importance was the appropriate use of symbolic elements, as "carelessness in the choice of words might result in death of the composer, or the person for whom it was composed" (Pukui 1949, 247).

Symbolic elements

Symbols maintain a repository of meanings, encapsulating our attempts to cognitively organize our spatial/temporal knowledge. However, they are useful only if people recognize them as standing for specific places or concepts. They can take material forms, such as an icon, a pattern, or a color, or immaterial forms, such as sounds, words, and gestures. In fact, every word is a symbol for some concept or relationship between concepts. Although they may be highly abstracted, frequently generalized, and may even change their meanings over time, symbols are essential to everyday communication, so much so that the meaning and value of any particular symbol may go well beyond the recognition of the identity of the thing or place it represents.

Symbols are capable of calling to mind a "succession of phenomena that are related analogically or metaphorically to each other" (Tuan 1990, 23). Within any particular culture, a symbol can act "as a trigger to help us recall the characteristics of that place, the specific whatness, whereness, and whenness information that gives it a unique identity. Given the symbol, we can fill in the necessary detail" (Downs and Stea 1977, 91–92). Kanaka Hawai'i developed complex symbolic systems, assigning certain attributes to specific things, such as types of animals, plants, or weather patterns. Although incorporating symbolic elements into any of the Hawai'i verbal arts is not a required feature, a *mea haku* who is skilled at doing so is highly valued (Donaghy 2011, 159). Four of the more common symbolic elements among those used by *mea haku* are complex metaphors, *kaona*, *oli* genres, and *wahi pana*.

Complex Metaphors

The use of metaphor in *mele* has been well documented since the mid-twentieth century (Pukui 1949; Elbert and Mahoe 1970; Tartar 1982;

Hoʻomanawanui 2005; Donaghy 2011). Metaphoric expressions are commonplace in Hawaiʻi verbal arts and are indicative of Kanaka Hawaiʻi perceptions of and interactions with the natural world. A skillful *mea haku* is capable of weaving together "metaphors of places and allusions to people using images of nature" (Stillman 2003). Pukui stated, "Persons were sometimes referred to as rains, winds, ferns, trees, birds, ships and so on" (1949, 248). In *mele aloha*, it is common to refer to a person as a flower or to describe lovemaking with images of water such as waterfalls, cresting waves, and streams. According to human geographer Paul Rodaway, "the use of metaphors to describe sensuous experience also reflects the basic multisensual character of geographical experience and the complex interrelationship of the senses." He further notes that some metaphors are so common, we forget they are metaphors, though they play a key role in expressing our sensuous engagement with the world (1994, 36).

The most obvious use of metaphor in *mele* are the *mele maʻi*, poetic expressions celebrating the organs and act of procreation. The words chosen for *mele maʻi* were purposefully figurative and playful, using features from the natural environment to illustrate the look, feel, or activity of the genitalia. One of the most well-known *mele maʻi* honors the genitalia of King Kalākaua.

"Kō Maʻi Hōʻeuʻeu"

Kō maʻi hōʻeuʻeu
Your animated aroused genitalia
Hōʻekepue ana ʻoe
That you are hiding
Hōʻike i ka mea nui
Show the big thing
O Hālala i ka nuku manu
Hālala, to the bird beak

ʻO ka hana ia o Hālala
What Hālala does
Ka hapapai kīkala
Raise the hips
Aʻe a ka lawe aʻe ʻoe
And take you
A i pono iho o Hālala
Right below, Hālala

Kō ma'i ho'olalahū
Your genitalia swells
I kai 'ale pūnana mele
(due to) sea swells a song nest
'O ka hope 'oi iho ai
And finally
A i pehu ai kō nuku
Your swollen mouth

Ua pā kī'aha paha
Take a drink perhaps
Ke noenoe mai nei
Foggy then
Ha'ina mai ka puana
Tell the refrain
'O Hālala i ka nuku manu
Hālala, to the bird beak
(Elbert and Mahoe 1970, 67; explanation expanded by Louis)

Mele ma'i are not considered lewd or risqué and are usually written for a specific high-born honoree while in infancy, in this case Kalākaua. This particular *mele ma'i* was written at a time when the Kanaka Hawai'i population was being severely reduced by the processes associated with colonization and was intended to inspire the Kalākaua to have many heirs.

The *mea haku* gave Kalākaua's *ma'i* the name Hālala, overly large, and instructs him in the first verse not to hide it but to show it off. The phrase in the last line, *i ka nuku manu*, is a metaphor for the community that gossips like squawking birds. In this case it is meant to describe the community's praise-filled chatter about Hālala. The second verse really leaves little for the imagination, but the third verse is a bit of a conundrum. After all, what exactly do swelling seas have to do with a song nest, unless the nest is a metaphor for the female genitalia that causes Hālala to become swollen—ultimately the best end, or place, *ka hope 'oi*, to satisfy its hunger. The last verse encourages Hālala to "take a drink" or dive in to experience a "foggy"-ness brought on by the intoxication of sexual release.

The intent behind the *mele ma'i*, to rejoice in the procession of a person's lineage, is not just a Kanaka Hawai'i value. A similar example of this would be the phrase "long live the . . .". The main difference is that Kanaka Hawai'i enjoyed complete openness about all things sexual prior to the arrival of missionaries and thought nothing wrong with the joyful veneration of fertile and powerful descendant-bearing genitalia. One can only imagine the horrified expressions on the faces of prudent-minded missionaries at the time this *mele* was composed and performed.

Kaona

Many people confuse *kaona* with metaphor, but the two are not the same. Metaphors are words or phrases meant to symbolically represent something that they are not. *Kaona*, on the other hand, represent the underlying reason some compositions were created in the first place, and are meant to "express a story in a manner that obfuscates the origins, characters, actions and meanings" (Donaghy 2011, 173). The *kaona* of a *mele* can be known only if the composer explicitly explains the meaning and story behind the composition, as Pukui shares with the *mele* "Ka ʻIwalani."

"Ka ʻIwalani"

Kaulana e ka holo a ka ʻIwalani
Well liked is the sailing of the ʻIwalani
Ke kaʻupu hehi ʻale a o ka moana
Moving like a sea eagle over the waves
ʻAole i ana iho koʻu makemake
Endless indeed is my admiration
I nā ʻiwaʻiwa o ka uka o Hāʻao
For the maidenhair ferns of Hāʻao
I ahona Honuʻapo i ka lau niu
Honuʻapo is made pleasant by the coconut leaves
I ka holu i ke ahe a ka makani
That sway with the wafting of the breeze
Aia i Punaluʻu, kaʻu aloha la
Over at Punaluʻu is the one I love
I ke kai kauhaʻa a ka malihini
Beside the dancing sea, the delight of visitors
Ke huli hoʻi nei o ka ʻIwalani
Now the ʻIwalani is on its homeward way
E ʻike i ke kai malino a o Kona
To the smooth sea of Kona
No Kona ka makani, he kulaʻi pau
To Kona belongs the gusty wind

Kīkiʻi kapakahi o ka ʻIwalani
That heels the ʻIwalani over to its side
(Pukui 1949, 248)

Pukui explains that though this *mele* appears to describe the voyage of the interisland steamer ship named *ʻIwalani*, it is really concerned with hiding the identity of the four women from Kaʻū, all of whom were related to Pukui's mother. Apparently the captain of the *ʻIwalani* was the stereotypical sailor, with a woman in every port. The "maidenhair ferns" were two women from Hāʻao. The woman from Honuʻapo was very tall, slim, and "frail enough for the wind to blow about," matching the image of a coconut tree whose leaves "sway with the wafting of the breeze" (1949, 248). Lastly, Pukui describes the woman from Punaluʻu as an older woman who still retained her beauty. Unfortunately for the captain, the woman he knew in Kona was not pleased after hearing of his exploits in Kaʻū and "stormed in her wrath, hence the wind that heeled the *ʻIwalani* over on its side" (1949, 248).

It is easy to understand why a composer may not want to divulge the underlying reason a *mele* is composed. To some degree, secretly expressing oneself is a very cathartic process, and maintaining the anonymity of characters or situations thwarts any backlash or squashes any chance at embarrassment. Furthermore, some intimately composed *mele* are "only intended for a particular individual—a lover, spouse, child, or beloved ancestor" (Donaghy 2011, 175).

Vocal Qualities and *Oli* Genres

Some Hawaiʻi vocal performances are aurally symbolic, in that the sound is indicative of the type of *mele* being performed. When I took *oli* classes from *kumu hula* and Hawaiʻi dance teacher John Keolamakaʻainana Lake, he explained there are at least four different vocal qualities a chanter develops over time, including *ʻiʻi*, *kāohi*, *hoʻānuʻunuʻu*, and *haʻanoʻu* or *aheahe*.

ʻIʻi—a greatly admired deep vocal tremor similar to a vibrato singer. It refers to the rasping articulation of consonants, h, k, the glottal stop, and vowel, u, o, a, especially after the former consonants. It is a quality used in varying degrees of emphasis and coloration. It is decidedly considered a most desired trait or skill of a chanter.

Kāohi—the cutting off of prolonged vowel sounds. It has a marked effect on pitch quality.

Hoʻānuʻunuʻu—quick trilling, pulse-like sounds that interrupt a continuous sound. The transition between pulses on a single pitch level causes partial slurring.

Haʻanoʻu or *aheahe*—loud, forceful articulation contrasted with soft, gentle sound.

(Lake n.d.)

The distinct vocal stylings used by an *oli* practitioner are associated with particular genres of *oli*. Stillman identified five distinct genres of *oli*: "*kepakepa,* rapid conversational patter, pitches not sustained, *kāwele*, more sustained declaration than *kepakepa,* but pitches still not sustained, *olioli,* recitation on sustained-pitch monotone, with embellishment using upper and lower neighbor tones, *hōʻaeʻae,* patterned and contoured use of sustained pitches, *hoʻouwēuwē*, funerary wailing" (2005, 79). Lake clarifies that the *ʻiʻi* is never used with *kepakepa*; *kāohi* is effective for *hōʻaeʻae* and *hoʻouwēuwē*; and *aheahe* is used for *olioli* and *hōʻaeʻae* (n.d.).

Furthermore, each *oli* genre is associated with specific types of *mele*, making both the vocal styles and the vocal performances aural symbols. *Kepakepa* is best suited to lengthy recitations, such as *mele inoa* or *koʻihonua*, because its rapid-fire delivery allows the performer the opportunity to complete the *mele* quickly, with fewer breaths. *Mele aloha* is usually performed using the *hōʻaeʻae* vocal style, with its intense use of *ʻiʻi* adding an emotionally powerful intrigue to the oration. Lastly, the *kanikau* uses a *hoʻouwēuwē* style, as it literally sounds like a person crying, so overwhelmed in pain that the observer/participant can feel the sorrow. (Lake n.d.)

Kanikau is one of the least heard chant forms these days, likely because of the reluctance to perform a lament when no actual death has occurred; or perhaps because of the personal nature of the *kanikau* and the emotional investment of the oral performance, it was never meant to be repeated. Nonetheless there was a time when it was not uncommon for professional wailers to attend a wake and offer *kanikau* for the deceased. "The art of the *kūmākena* was, perhaps, a professional and public oratorical form of sending a person into the next world with social approval and stylized bereavement. We would assume that they were also rewarded with some sort of *hoʻokupu* [contribution] after the event" (Johnson n.d.). However, this vocal performance is making a comeback. I had the privilege of hearing a *kanikau* performed at my godfather's services.

According to Arista, *kanikau* made a highly successful transition "into printed literature by virtue of their frequent appearance in Hawaiian language newspapers throughout the 19th century" (2015). She describes these texts as compositions "chanted to honor the deceased and send the soul on its final journey to dwell with ancestors in the spirit realm. These chants—from 25 lines up to 250 lines long—are the expressions of lovers, relatives, friends, and admirers reflecting on shared experiences of life, work, travel, the rearing of children, love, and more" (2015). *Kanikau* were written and included in Hawaiʻi language newspapers daily, providing access to a larger public audience. As painful as the passing of a loved one is, *kanikau* were also meant as poetic creations of love, praise, and passion. The following is a second-year Hawaiʻi language translation assignment given to me by Kuʻulei Kanahele from Hawaiʻi Community College.

"He Kanikau Aloha Keia nou e Kiope"

Kuʻu wahine i ka uluniu o Mokuola ē,
My beloved wife of the coconut grove of Mokuola
Mai ka iʻa haʻaheo lā ma ka moana,
From the proud fish of the ocean
Ke huki kū maila i ka ʻilikai o Hilo One, aloha wale,
Pulled up to me from the sea at Hilo One, only love,
Kuʻu wahine i ke one loa o Punahoa ē!
My sweetheart of the long sand of Punahoa!
Aloha ia one a kāua e hoʻolaʻi ai ma luna o ka lio
who loved her homeland as we sat peacefully poised on the horse

Kuʻu wahine o ke ahiahi Pōʻaono o ke Kaona ē
My sweetheart [born] of the 6th night of Kaona
Mai ka Pele kuʻi lua lā o Haili,
from the twice struck bell of Haili,
Mai ka Piana kani hone lā i ka ʻiuʻiu,
from the sweet sound piano in the far off distance
Kuʻu wahine i ka uka anuanu nāhelehele, o Kumukukui ē
my sweetheart of the cold undergrowth of Kumukukui
ʻO ia wahi a kāua i alo ai i ka hau kololio o ke aumoe
In this place together we endure the cold gust of night

Kuʻu wahine leilehua o Mokaulele ē!
My sweetheart, a lehua garland of Mokaulele!

Ke hoʻopulu ʻia aʻela e ka ua liʻiliʻi o Makahuna,
Saturated with the small rain of Makahuna,
Huna nā maka o kuʻu hoa lā nalowale loa,
I will no longer see my beloved friend long lost,
Pau ka pili ʻana o ke kāne me ka wahine aloha,
our connection as man and woman is ended,
Ua haʻalele mai nei ʻoe iaʻu me ke keiki a kāua,
you left me with our child,
ʻO kaʻu mea nui nō ia o ke kaikamahine aloha
Our daughter's love is now my most cherished love

He hōkū hāʻule ʻole ʻia naʻu i ka wā e ola ana ʻoe,
Your star never fell to me in the time while you lived,
Kuʻu wahine i ke kula loa o Puainakō ē! koʻu hoa,
my sweetheart of the long field of Puainakō! my friend,
Kuʻu wahine i ka ʻōhiʻa loloa o Honokowailani,
my sweetheart of the long ʻōhiʻa of Honokowailani.
Mai ka ua noe hāliʻi mai i ka nahele ē,
from the misty rain covering the wet forest
Kuʻu hoapili o ka ʻāina ua nui o Hilo nei;
my dear close friend of the big rain land of Hilo,

Mai ka ua ʻaweʻawe lua lā ma ka moana,
From the two tentacle rain of the ocean
Mai ka ua nihi hele aʻe ma kai o Leleiwi ē,
from the rain carefully moving up toward the ocean of Leleiwi,
Kuʻu wahine i Ka ua kanilehua o Panaʻewa,
my sweetheart in the Kanilehua rain of Panaʻewa,
Ke lei haʻaheo maila i ka liko o ka Lehua,
The proud lei from the bud of the Lehua,
ʻAuwē kuʻu wahine ē! kua hoapili o ka ʻāina ē!!
ʻAuē my sweetheart! My dear close friend of the land!!
J. W. K. Wahineole K. Kalepolepo, Hilo, Iulai [July] 14, 1868.
Ke Kuokoa, Honolulu, Augate 1, 1868

In this lament, the *mea haku* shares personal memories of a dearly loved woman. The *mea haku* begins by naming what could possibly be the place where they met, on the small islet of Mokuola of the shore of Hilo Bay, and then likens the woman to a proud fish who is drawn by the attraction of love. The *mea haku* remembers how the woman loved her homeland, an area known as Punahoa near the shoreline of Hilo Bay, as she sat on the back of a horse. The second verse could be a memory of their wedding, as Haili is the location of the Catholic church. Perhaps they spent a wet, cold evening in Kumukukui for their honeymoon, where they were warmed by their passion for each other. In the third verse, the *mea haku* laments the loss of the woman and ensures the daughter she left behind will be cherished. In the fourth verse, the *mea haku* affirms that the woman never fell from the pedestal she was placed on and lists several places where they possibly spent time together. In the last verse, the *mea haku* uses the imagery of rain to describe the depth of sorrow felt for the loss of this woman.

Throughout the *kanikau*, the *mea haku* uses several place and rain names. Arista notes this is a common element in many *kanikau* and further states that "many of these places and the names attached to them have been nearly forgotten, and chants like the *kanikau* are a kind of place repository through which Hawaiians may reacquaint themselves with their homeland. Lengthy chants list 50 or more different places, not all of which are identified on contemporary maps or in reference books" (2015).

Hawaiʻi place names provoke, reveal, and provide attachments to the land, to the past, and to those pivotal events a person holds dear. They are often understandable only with the *moʻolelo* that accompanies them and usually only by those who have insider knowledge from the *mea haku*. Knowledge of their meaning provides insight on the importance these place names had in shaping the lives of Kanaka Hawaiʻi. Sharing the names and meanings of places was a conscious act of cultural regeneration. Kanaka Hawaiʻi incorporated themselves and their experiences into the islandscape using place names as storied symbols in their cartographic tradition.

Wahi Pana

Wahi pana are a special set of place names that represent entities and situations of consequence purposefully woven into the islandscape to serve as constant reminders of past events, cautionary tales, and epic tragedies. They are a symbolic shorthand applied to various features for the purposes of finding them again, referring to them in casual conversation and rituals, and passing on the knowledge of "what happened here" to other people. Understanding how and why Kanaka Hawaiʻi named places and how and who decides which place names are remembered is a complicated endeavor, because it is difficult for some people to separate themselves from their connection to the places, times, events, and people some place names represent, thereby making them "arguably among the most highly charged and richly evocative of all linguistic symbols" (Basso 1996, 76–77).

Wahi pana developed organically and became coherent local mnemonic and symbolic devices that unfolded the richness of the Kanaka Hawaiʻi conscious connection with and understanding of the world. Naming anything creates a connection, making it familiar and allowing people to bring it into their consciousness. *Wahi pana* are found in all Hawaiʻi verbal arts, poetic and nonpoetic, and can be used either metaphorically or as *kaona* to enhance the depth of a composition. According to American anthropologist and Polynesian folklorist Katherine Luomala, "Hawaiians . . . incorporated numerous place names into their narratives. A composer selects place names for his poems which express his emotion in regard to another individual to whom he dedicates his chants" (1965, 238). A highly skilled *mea haku* would weave in as many place names of an area as possible as "witness both to the story's veracity and the teller's memory" (Pukui et al. 1974, 272). According to Hoʻomanawanui, "the mere mention of a specific place name recalls to the educated mind the countless number of *moʻolelo* associated with that place, preserving both the name and the story of the area" (2013, 164).

The opening *mele* for this chapter, "Hilo Hanakahi," as well as the one for the book, "Kūnihi ka Mauna," demonstrate how *wahi pana* are used poetically; the first preserves the name associated with each place and the second preserves the *moʻolelo*. "Hilo Hanakahi" is an example of how *wahi pana* are not just terrestrial, as winds and rains are also given names that are recognized and celebrated in *mele*. "Kūnihi ka Mauna" is an example of how a Kanaka Hawaiʻi social etiquette incorporated into a *moʻolelo* becomes associated with the *wahi pana* from the *moʻolelo*. Whether descriptive or commemorative, *wahi pana* are situating devices that spatially anchor and locate narrated events, rendering the islandscape intelligible (Basso 1996, 40–47). To know and speak the names of the places where one lives and travels is to be reminded of the events that happened at those places and to be connected to those ancestors. Weaving those place names together in a story allows information about the location of food and shelter to be recalled and reintegrated into the communal knowledge base.

Training to be a *mea haku* involves so much more than what has been presented here; this brief summary is meant only to express the diversity of topics and training techniques necessary for a *mea haku* to learn. Skilled orators were and still are highly valued in Hawaiʻi. Many consult with each other when they compose new *mele* to ensure they use the mnemonic and symbolic devices appropriately, as they are all too aware of Pukui's warning about a careless choice of words.

The next section focuses on a particular type of performance, the *haʻi moʻolelo*, Hawaiʻi storytelling, which does not require extensive training but does require the *haku moʻolelo*, Hawaiʻi composer or composition of historical records, to intimately know the topic. The topic presented as an example is the storied *wahi pana* of Kealakekua. While I have not spent a lifetime in the Kealakekua area of Hawaiʻi Island, I did sit with and learn from Aunty Moana Kahele, a Hawaiʻi cultural practitioner and community-recognized respected elder. She believed that if we stop telling our stories about the places where we live, work, and pray and supplant them with someone else's stories, then we sever our cultural connection with our ancestors. Telling stories helps people maintain those cultural connections. The stories in the following section belong to her family. They have graciously allowed me to share them with you in this book.

Ka Haʻi Moʻolelo ʻAna (Storytelling)

One of the ways Kanaka Hawaiʻi express their symbolic understanding of nature is through the continuous retelling of *moʻolelo. Moʻolelo* should not be confused with story. It is not fanciful fairy tales or fictional stories woven to entertain. It is an invaluable source of information laden with the skillful use of mnemonic and symbolic devices meant to preserve Kanaka Hawaiʻi consciousness and connection with a multidimensional reality. There were hundreds, if not thousands, of *wahi pana*, many of which are no longer part of the Kanaka Hawaiʻi consciousness because too many people are more interested in high-definition action or fantasy that provides a convenient escape from everyday reality. Hawaiʻi names were not randomly given. According to Aunty Moana, there was a reason for a place being named; such is the case for Kepuhi and Limukoko.

Kepuhi is the blowhole of the Kealakekua area that acts as an early warning system for storms. Depending on the time of year, women and young children would go to the shoreline to wash clothes, and some still go to gather food. After years of observation, they learned to listen to the way the waves sounded crashing at Kepuhi. If you hear only the water rushing through the rocks, it is okay for the fishermen to go out for deep-sea fishing. However, if it sounds like an underwater explosion of rocks slamming into each other, it is time to bring the canoes in and lash them down, because rough weather is approaching.

Limukoko is a rock near the shoreline by Palemanō that had a lot of *limu kohu*, a variety of seaweed. Anybody could go and pick as long as they were watchful not to stay too long and overpick. If you were not mindful, the water would rise and make it hard, if not impossible, for you to get back to shore. According to Aunty Moana, the ocean had a way of making sure you took only what you needed and didn't get greedy.

This *moʻolelo* teaches the next generation to be aware of their environment and warns of the consequences of being greedy. The people living near the shoreline of the Kealakekua area paid attention to the sounds and cycles of their natural environment. They knew powerful ocean currents predict the arrival of those atmospheric events that created them. So when they heard rocks slamming into each other at the coastline, they knew it was a warning. They also knew the tides provided only a small window of time to safely pick *limu*, and that the amount of *limu* you can pick in that time limit is enough to feed your family. Picking *limu* for a longer period of time will put you in harm's way, and you risk losing it all to the ocean.

Another *moʻolelo* Aunty Moana shared that reminds people about the consequences of being greedy relates to the name Nāpoʻopoʻo. Her handwritten account follows.

Legend of Napo'opo'o

{By my Paternal Grandmother - Mary K. Kapule & father
Isaac K. Kapule Sr.
Written by Mona K. Kapule Kahele - 1937 - [m] Collection}

Kai-a-ke-kua had two great ponds. The ponds were clear and fresh pure water. Pond number one was at Keei and shallow. Pond number two at Kai-a-ke-kua. These ponds were circle, except for pond number two was shape like a bowl or basin.

The people never went without water. They had more water than they can use. Everything was just dandy. Lots to eat and lots of good water to drink. The people were happy. Suddenly, there was trouble brewing between the people and the aliis. (Royalties)

The aliis wanted the water all for themselves, and this enrage the people so much. All their lifetime, they depended on these ponds and water for their uses too, and so as the royalties.

When a new alii became the chief, he was mean. He treated the people so bad that they cried out to their gods for help. Before they knew what was happening this chief ordered the other aliis to place a kapu on these ponds. The people were told they could not have any water from these ponds because it belongs to the aliis. The people suffered very much. They had to drink the salt sea water from the beach. No fresh water to get for miles and miles. Death was the penalty if they were caught taking water from the ponds. The troubles went on and on. More people were been killed for taking water and the kapu system became more and more worst. If one child is caught drinking from the pond that child and the entire family were killed. How greusome can these alii be.

Then one morning when the people and the aliis woke up, there were no ponds of water to be seen. They just seem to disappear. In its place, the rocks were just as dry as can be. It seems like no water ponds were there. So strange.

The aliis went to the Kahuna to ask or tell them what happened to the ponds of water. The Kahuna answered, "The ponds had gone away because you are so mean and greedy". You made the people suffer and so the ponds made you to suffer too. Where the ponds are I cannot tell you, for I do not know.

Strange as it seems the pond at Keei Beach moved up to the higher slopes and became a great swamp and known as Wailapa. The other went to the ocean and became Wai-lapa-kai.

However the story is, they were brother and sister who had come from the Hidden Island. They traveled until they landed at Kai-a-ke-kua. They looked for water to drink, but there was not any around. All they could see was barren land for miles and miles. No trees or grass.

2

Not, even the birds or anything on land.
So keen children of kupuas, they changed
themselves to ponds of water.
The birds began to arrive. Trees and
grass began to grow. Pretty soon most of the
land was green with all kinds of plants.
People began to come. They lived happily
for a long time until a mean Alii lived
with them and than troubles began.
The meanness and greediness of the aliis
made the people suffer. So in the place it
left a great dent. This is how Napoʻopoʻo
was named.
Na poʻopoʻo = means the dent.
"The End"

The *moʻolelo* has to do with the mismanagement of a sacred relationship that Kanaka Hawaiʻi from this area maintained with water. In Hawaiʻi tradition, the *konohiki* is a land steward responsible for managing the respectful relationships the people of the *ahupuaʻa* nurtured with natural entities. This story is a warning for anyone in a leadership role to responsibly manage all interactions with those relationships because they are "gifts from the gods." Any kind of mismanagement, be it greed or irresponsible behavior, will lead to dire consequences for all. Climate change is an excellent example of the consequences of irresponsible behavior by those people in positions of power.

In the *moʻolelo* associated with the name Kealakekua, there are also good examples of responsible behavior being rewarded. Here is another of Aunty Moana's handwritten texts.

Kealake-kua (Pathway of Kua) collection

(Told by Uncle Henry Leslie Sr. I Written by Mona K. Kahele)

Many people translate Ke ala ke kua as the pathway of the God. But however the questions arise as to what god? There are quite a number of stories were written about Ke ala ke kua. But this is the story I heard from one of my uncles, who lived in the village when he was a young man. He heard this story from one of the kupunas of the village. So here is the story.

First Kealake Kua, not as Ke ala ke'akua as sometimes heard among some people.

Long long time ago the Hawaiian people had lots of gods. The earth, heavens, stars, moon, land, sea, rocks birds houses, pigs, chickens, trees, grasses, canoes and man. In fact everything of nature had a god. But they kept the four main gods. Kane, Ku, Lono and Kanaloa. To-day, known for tradition as the Four Major Gods for their Culture.

However, the people also had a god and a Great King of all sharks in the ocean. This shark god was known as Kua.

He was also the King Shark of Kau and the ancestor of some people of Kau and Kona. At times he can become a man without anyone suspecting anything. He does not bear any marks of any kind.

Kua traveled the waters from Kau to Kona and around Hawaii and the other islands. On his return trip to Kau as usual his stops were always at Kapukapu, Kona for a nice rest before returning to Kau.

On one of his trips he decided to spend sometime and mingle with the people of Kapukapu. Changing himself completely to a man he mingled with the people. The people recieved him with aloha and treated him well. He lived among the farmers. Did everything what a makaainana (commonier) did. From planting, fishing, building canoes and etc. Things he never done in his life he was doing it now.

The fishermen loved him the most because he was good at catching fish. Anyway was very cooperative and showed all his aloha to the people. Likewise the people treated him so well and showed their gratitude towards Kua. They didn't treat Kua lik a visitor but as one of them.

Finally the time came when it was time to tell these wonderful people that he must leave them and return to his people. Upon this news, the people were so sad. They felt like a broken family. They beg him to stay on with them, but he thanked them and said, I will come to visit you some day. I had enjoyed my stay here. You treated me like a relative and not as a visitor Let me leave you these words, when the sun is golden, the horizon red and gold I will be here. However you will be safe in these waters or wherever and even your Generations to come.

The people did not reali[illegible] or any feeling that Kua was King of the Sharks

Kua stepped into the water cupping his hands over his mouth, he bent down and gave a booming voice like a chant only he knows.

Suddenly, there were two dark lines on the water. It formed two straight lines from the horizon to the shore. Kua stood still and watch. Even the people on land were all silent. They were wondering and whispering, what kind of man this is? He chanted over the water and there are two dark lines coming to the shore. Must be he is an alii and all the canoes are coming. But, as the dark lines touch the shore the peoples voices sounded "alas this must be a god."

Kua stood in the center of the two dark lines, he turned around and waved to the people. By this time the bottom part of his body was changing. Not saying a word, he dived into the water. When he appeared he was a huge shark swimming between the two lines of sharks. All swimming out towards the horizon as the sun was setting in the ocean.

Now the people were sure, Kua was none other but the great King of the sharks. They were so happy that they had treated Kua with all their kindness and aloha.

For all the good deeds Kua did for the people where he stepped into the sea, the people named that spot Kai-a-ke-Kua (waters of Kua) and thats where the wharf stands at Napo'o-po'o. The trails of pathways where he walked was named Ke ala ke kua (pathway of Kua). As far as anyone been bitten by sharks at Napo'opo'o, none that anyone knows or heard about.

This is how Ke ala ke Kua existed. Today,

we have a Post Office by that name. Banks
stores, hospital Medical Center, flower shops
and numerous other important buildings
are located in this area named,
"Keala kekua".

The End.—

The *mo'olelo* of Kealakekua teaches future fisherman that the bay is a protected fishing ground. Fishermen need not fear seeing *manō*, shark, swimming within the bay. As long as they continue to care for their relationships with respect, there will never be a shark attack within the bay. Of course this inspires the question of what the possibility of an attack in the future would mean to the story. How or would the *mo'olelo* still have relevance? When I asked Aunty Moana this question, she said that if anybody gets attacked in the bay, it would be because they didn't show respect in one way or another.

Not showing respect could happen to greedy fishermen or careless swimmers. Maintaining belief in the old way is a sign of respect as well. All the fishermen I spoke with from this area said they frequently see sharks but they don't bother each other because they are not taking what they do not need to feed their families. One young fisherman even said there is a huge shark roaming around in the area but that it never really comes into the bay, it just stays near the ledge at Palemanō. He sees it regularly when he goes diving but never fears being attacked because he knows he is safe fishing in the bay.

Kealakekua is an important *wahi pana* that also relates to the start of Makahiki season. It is most commonly known from the perspective of *nā ali'i*, Hawai'i political leadership class. It is considered the starting point for "the path of the god"; Lonomakua as a representative effigy is carried by *nā kahuna*, Hawai'i class of master practitioners, who accompany *nā ali'i* on a clockwise procession around the island, collecting tribute at the stone altars of each *ahupua'a*, the land area named because of the stone altar on which residents would place offerings as part of the Makahiki celebrations. The fact that there are two *moʻolelo* for one *wahi pana* does not make one more right than the other; it just makes them contextually specific. Whether it is a god living and working among the people or a reenactment of an ancestor elevated to *akua* status through generations of ritualized performances, both *moʻolelo* associated with Kealakekua acknowledge and weave the presence of divine entities into the Hawaiʻi landscape.

Perceptions of Kanaka Hawaiʻi Spatial/Temporal Knowledge Symbolization

From the earliest time, the performance of Hawaiʻi verbal arts has served several interlocking purposes: the celebration of the beauty of the islands, the birth of the universe, the loss of loved ones, the playfully direct propagation of a lineage, and the purposefully hidden messages of sexual exploits. *Mea haku* were not confined to *nā aliʻi* or *nā kahuna*. *Makaʻāinana*, Hawaiʻi general populace, also "wove living threads from their own historic circumstances and everyday experiences into the ongoing oral tradition, as handed down from expert to pupil, or from elder to descendant, generation after generation" (Pukui and Korn 1979, back cover).

According to Pukui, the most talented *mea haku* were "versed in poetry (mele), storytelling (haʻi kaʻao), genealogy (moʻokūʻauhau), and oratory (kākāʻōlelo) and found themselves in the courts of the chiefs" (n. d.). In these settings, orators would often include *oli*, *ʻōlelo noʻeau*, *ʻōlelo nane*, and gestures as part of the performance. *Mea haku* encoded ancestral knowledge for future generations using several mnemonic and symbolic devices. They trained worthy apprentices who were selected at an early age using techniques that involved attentive listening, repetitive recitations, and breathing practices. These methods ensured Kanaka Hawaiʻi spatial/temporal knowledge was accurately maintained for future generations and any new compositions were completed with the same level of rigor and attention to the appropriate use of words.

Kanaka Hawaiʻi enjoy playing with words, "selecting and arranging them, to achieve patterns aesthetically pleasing to themselves" (Luomala 1965, 234). *Mea haku* cultivated symbolic meanings for the sounds and incorporated multiple levels of meaning in their various compositions. They also recognized *wahi pana* as repositories of human experiences that serve "as durable symbols of distant events and as indispensable aids for remembering and imagining them" (Basso 1996, 6). *Wahi pana* and their cultural meanings are generated by the constant interaction of people and their environment. Ultimately, Hawaiʻi verbal arts are a manifestation of participating with the natural environment, the embodiment of experiencing the world sensuously and imbuing it with metaphoric meaning.

CHAPTER SIX

Ka Hula

(Hawaiʻi Dance)

"Ke Haʻa lā Puna i ka Makani"

Ke haʻa lā Puna i ka makani
Puna dances in the wind
Haʻa ka ulu hala i Keaʻau
Moving through the hala grove at Keaʻau
Haʻa Hāʻena me Hōpoe
Hāʻena and Hōpoe dance
Haʻa ka wahine
The female sways
ʻAmi i kai o Nanahuki
Revolving at the sea of Nanahuki
Hula leʻa wale
Perfectly pleasing, the dancing
I kai o Nanahuki
At the sea of Nanahuki
ʻO Puna kai kuwā i ka hala
Puna's sea resounds in the hala
Paepae ka leo o ke kai
The voice of the sea is carried
Ke lū lā i nā pua lehua
The lehua blossoms are scattered
Nānā i kai o Hōpoe
Look toward the sea of Hōpoe
Ka wahine ʻami i kai o Nanahuki
The dancing woman at the sea of Nanahuki
Hula leʻa wale
Perfectly pleasing, dancing
I kai o Nanahuki
At the sea of Nanahuki
(Kanahele 2011, 110–111)

Ka hula, Hawai'i dance, is a performance-based encapsulation and transmission of a Kanaka Hawai'i sensual engagement with and perception of nature, bringing choreographed rhythmic movement to symbolically encoded compositions. A *kumu hula*, Hawai'i dance teacher, interprets the *mele*, Hawai'i song, chant, or poem composition, rife with mnemonic and symbolic devices, adding even more layers of meaning through choreographed movements, rhythmic accompaniment, and adornments, each of which add some kind of symbolic meaning. The above *mele hula*, a Hawai'i composition meant to accompany *hula*, is part of the Pele family tradition documenting the origins of the *hula Pele*, the Hawai'i dance genre that honors Pele.

According to Emerson, the *mo'olelo*, Hawai'i historical account, associated with this *mele hula* marks the start of an epic journey, a saga that extends across the Islands of Hawai'i. While on a family outing, the Pele *'ohana*, family, traveled from their home at Kīlauea to Nanahuki. Pele was pleased at the sight of Hōpoe dancing on the beach and asked her sisters to reciprocate the entertainment with their own dance. Hi'iaka was the only one who responded favorably to the request, composing and then dancing to the opening *mele hula* (Emerson 1997, 1–2). Pele has several Hi'iaka sisters. Generally speaking, if a storyteller uses the name Hi'iaka, it refers to Hi'iakaikapoliopele, the youngest sibling born in the shape of an egg, full of potential and possibilities. This *mele inoa* starts Hi'iaka on a journey that ultimately ends with her transformation into an elemental entity associated with generating life on the barren lava fields.

The composition includes specific place names, recording the location as Nanahuki, a beach in the *'ili*, land division term, of Hā'ena; located in the *ahupua'a*, land division term, of Kea'au; within the *moku*, land division term, of Puna (Kanahele 2011, 112). It also includes *mahuahua*, reduplication or a repetition of syllables, a literary device that helps a performer easily recount the *mele hula*. In this case, the use of the term *ha'a*, Hawai'i dance with bent knees, in the first four lines, and the phrase *i kai o Nanahuki*, at the sea of Nanahuki, helps both orator and observer/participant easily situate both the event and the place it occurs, dancing with bent knees at the sea of Nanahuki. The composition also includes sensory elements distinctive to this particular place, which not only indicates the composer's familiarity with the place but also makes the composition more believable and easily accessible to the masses, such as the wind moving through the *ulu hala*, pandanus groves; the scattered *pua lehua*, *'ōhi'a* tree blossoms; and the sound of Puna's sea being carried through the *hala* groves.

A *kumu hula* makes use of the imagery created by the description, brings forth the resounding sea using the rhythmic beating of percussion instruments, and choreographs simple yet powerful movements simulating "the ocean currents at Hāʻena, the wind upon the grass and in the trees at Keaʻau, and the undulating tides of Nanahuki" (Kanahele 2001, 7). In the stage performance of "Holo Mai Pele," taped as part of the PBS series Great Performances and aired in 2001, a gracefully powerful, disciplined dancer performs this *hula* according to a choreography that has "always been done that way, passed on from generation to generation by our grandmother, our oldest cousin, our mother, and their teachers before them" (Kanahele 2001, 8). Knowing the movements being performed today are exactly the same as the day it was choreographed *i ka wā kahiko Hawaiʻi*, in old time or pre-contact Hawaiʻi, brings stability to the spatial/temporal knowledge being transmitted.

The *hula Pele* maintained by Hālau o Kekuhi is only one of several *hula* traditions. Other *hula* traditions maintain their own versions on the origins of *hula,* including the traditions of Laka, Hinauluʻōhiʻa, Moʻikeha and Laʻa, and Laʻilaʻi. There's even a suggestion that the *hula* was born out of the Hawaiʻi martial arts, *kuʻialua* (Kanahele 1986, 129). As noted in previous chapters, it is neither confusing nor uncommon for multiple versions to exist. Each is contextually relevant to the people and place from which it originates. The existence of multiple versions is an affirmation of the distinct *hula* styles that continue to be practiced today at an international scale, having proliferated into countries such as Japan, Mexico, and even several European nations.

Many preparations are necessary to become a *hula* dancer, and the sections that follow describe some of the spatial/temporal knowledge dancers must learn to embody to transmit this knowledge through a *hula* performance. The dancer's attentiveness in adhering to those Kanaka Hawaiʻi cartographic processes embedded in the performance sensually effects the observer/participant, possibly leading to an enlightened spatial/temporal awareness never before experienced. Much of the information shared in this chapter is derived from texts and the author's experiences, both as an observer/participant and as a student at Hawaiʻi Community College, where the author learned first-hand that to be a dancer requires no small commitment.

Preparing the Dancer

As *kumu hula* and Hawaiʻi mythology scholar Taupoui Tangarō explained during a beginning *hula* class I was enrolled in at Hawaiʻi Community College, the opening *mele hula* is evidence of the tradition of "teaching hula." Today *hula* is taught in many places, but not all of these places can be considered *hālau hula*, a Hawaiʻi place of learning *hula*. Should a person have the privilege of joining a *hālau*, staying in it requires humility and respect for all aspects of the *hālau*, dedication and strength of character to endure the rigorous immersive training, and patience and dexterity to create both the musical instruments/implements and the accompanying regalia.

Aspects of *hālau hula*

Hālau hula are places where *kumu hula* transmit the spatial/temporal knowledge they maintain according to their formal training in a specific *hula* genealogy. As *hula* evolved from a highly ritualized practice to a more palatable form of general entertainment, so did certain aspects of *hālau*. Today, while some *kumu hula* teach in *hālau*, others teach *hula* in dance studios or academies. Regardless of what a *kumu hula* chooses to call the place where they maintain and transmit Kanaka Hawaiʻi spatial/temporal knowledge, most *kumu hula* maintain a similar organizational structure.

Organizational Structure

The *kumu hula* is the undisputed head of the *hālau*, whose responsibilities are to ensure that traditions are maintained and transmitted to the next generation. Not all *hula* teachers are *kumu hula*. From a purist's point of view, a *kumu hula* is an individual who has studied extensively, in particular *hula* genealogy, and has gone through a ceremony known as *ʻuniki*, graduation exercises, where knowledge is bound to the student. A *kumu hula* is capable of telling you where their *hula* lineage(s) originates. However, not all *kumu hula* have the opportunity to go through a formal *ʻuniki* process. Nowadays, an *ʻuniki* can be as simple as a *kumu hula* telling an advanced student to go teach. According to *kumu hula* and Hawaiʻi scholar Kealiʻikanakaʻoleohaililani, a *kumu hula* must maintain and share the spatial/temporal knowledge they learned from their formal training in a particular *hula* tradition and must be acknowledged by other respected *kumu hula*.

All *hula* students are referred to as *haumana*. In some *hālau*, more advanced *haumana* are considered *ʻōlapa*, dancers with a natural grace and beauty who emulate the distinct style of the *hālau* and are capable of being soloists or being featured dancers when the group performs (Stagner 2011, 63). According to Hawaiʻi biologist and *oli* practitioner Sam ʻOhu Gon III, the word *ʻōlapa* also names one of the "plants only found in the *wao akua* whose leaves are in motion with the slightest breeze while nothing else in the forest is moving, drawing attention to the singular beauty of the tree" (2010). Associating a standing hula dancer with the *ʻōlapa* in the *wao akua*, the Hawaiʻi forest region for the divine entities (*akua*) associated with life-giving natural processes, metaphorically relates the graceful standing movement of the *hula* dancer with that of the *wao akua* forest dancer. An *alakaʻi,* leader, is the most accomplished *ʻōlapa*, who serves as an example for other *haumana* and is capable of leading classes at the request of the *kumu hula*. Kealiʻikanakaʻoleohaililani explains that this is a relatively new role within the *hālau*; it is not a term found in the older texts but is consistent with *kumu hula* adjustments to the demands of a different socioeconomic system (2016). The *hoʻopaʻa* are advanced *haumana* who have mastered and committed chants, dance, and rhythmic beats to memory and are capable of assisting in *oli* and instrumental accompaniments during classes at the request of the *kumu hula*. The *alakaʻi* and *hoʻopaʻa* give the *kumu hula* "a break from hours of instructing as well as chanting" (Stagner 2011, 64). Lastly, the *poʻopuaʻa* is an administrative assistant or spokesperson who takes care of scheduling events, oversees preparations for instruments and regalia, and, most important, tends the *kuahu,* altar (Emerson 1998, 26–29).

Kuahu

The *kuahu* serves as "the visible temporary abode of the deity, whose presence was at once the inspiration of the performance and the luck-bringer for the enterprise—a rustic frame embowered with greenery" (Emerson 1998, 15). It is the physical manifestation of the *wao akua* where Laka, a divine entity associated with *hula*, is metaphorically brought in to the *hālau* using greenery considered to be her *kinolau*, body forms. Some of the more commonly used plants include *lama*, an endemic ebony hardwood tree; *lehua*, flower of the *ʻōhiʻa*, a hardwood tree in the Hawaiʻi forest; *maile*, a native twining shrub with shiny leaves; and *palapalai*, a native fern. According to Gon, each has symbolic significance.

The *lama* wood is usually a centerpiece on the *kuahu* and is understood as enlightenment, providing *haumana* with a subconscious prompt to remain open to the teaching so the process can enlighten them. The *lehua* is a flower of the *ʻōhiʻa* tree and also describes an expert of any particular profession, inspiring *haumana* to practice with the purpose of perfecting their skills. *Maile* is a vine with shiny leaves and a refreshing fragrance that both fills a room and lingers, much like a *hula* performance fills and lingers in the memory of the observer/participant or a lesson taught in *hālau* is expected to fill and linger in the *haumana* body/mind memory. *Palapalai* is a delicate-looking fern that is surprisingly resilient, encouraging *haumana* to imitate its delicate movements while increasing their body's resilience and strength (Gon 2010).

Tending to the *kuahu* requires the utmost respect and knowledge of proper protocols. After all, the caretaker must enter the *wao akua* to gather and remove the greenery to be used, transport it to the *hālau,* and arrange it on the *kuahu*. Each step in this process requires *pule* informing the divine entity of the petitioner's intent and requesting permission to proceed. Simply entering the *wao akua* requires specific ceremonial protocols, as it is considered profoundly sacred and unassumingly perilous, as too many unaware travelers have learned after getting lost or, in the extreme, losing their lives or requiring state-funded rescues. *Kumu hula* are trained specialists with knowledge of how to access the *wao akua* using specific purposeful protocols (Gon 2010).

Hālau Hula Protocols

Hālau hula protocols are very specific, and *haumana* are expected to perform them according to the tradition each *hālau* dictates without question. These protocols vary depending on the *hālau*; however almost all of them require the *haumana* to perform *oli kāhea*, the chant performed to enter, prior to the start of training. It signifies the *haumana* is ready to submit to the learning process and will adhere to the *hālau* protocols. The further a *haumana* progresses in the *hālau*, the more protocols there are to follow. For example, I learned as a *haumana* in the beginning *hula* class at Hawaiʻi Community College that it is considered rude to "point" the bottoms of the feet at another person. So, when sitting on the floor, we had to sit either with both legs crossed or to one side. We also had to ensure that neither our apparel nor our body odors offended others. These are rather mild protocols compared with what the late *kumu hula* John Lake went through for his *ʻuniki*. He told our beginning chant class he had to abstain from all sexual activity for a year, which was difficult, as he was a married man. These are the kinds of sacrifices a person committed to a life of *hula* embraces for the love of their craft.

Types of *Hula*

Each *hālau hula* has its own style with which it executes different types of *hula* based on traditional teachings. Some *hālau* specialize in particular types of *hula*, such as Hālau o Kekuhi, which is known for its mastery of the *ʻaihaʻa*, a bent-knee dance style accompanied by emphatic and bombastic chanting associated with the eruptive volcanic persona of Pele. Another *hālau* with a notable style is Hālau Mōhala ʻIlima, which is known for its performances of tightly clustered formations of *ʻōlapa* moving as a single unit. The point is that *hula* "style" is not the same as *hula* "type." Two *hālau* might perform the exact same *mele hula* and it will look completely different, as each performance reflects the traditions of their specific *hālau*. Just as there are different styles of *hula*, there are also different types of *hula*, and many different classification schemes.

Some *hula* are performed while standing, others are performed while sitting; some dances use instruments, some are accompanied only by *oli*; some are deeply ritualistic and others are merely for entertainment. Most people separate *hula* into two broad categories: *hula kahiko,* Hawaiʻi dances reflecting more traditional rhythms, movements, and adornments; and *hula ʻauana,* Hawaiʻi dances reflecting more modern rhythms, movements, and adornments. Within these broad categories, Emerson and Beamer published their understandings on specific types of *hula.*

Emerson lists twenty-eight different types of hula in his book, *Unwritten Literature of Hawaii: The Sacred Songs of the Hula,* three of which are more games than types of *hula*. The rest can be loosely categorized according to the instruments used, the emphasis on particular body position or movement, the animals being revered, and the ancestors being celebrated. For example, the *hula pahu*, which features the use of a drum, and the *hula ʻiliʻili*, which features the use of smooth, flat hand-sized stones or pebbles, are types of *hula* based on the use of an instrument. The *hula paʻi umauma*, a chest-slapping *hula*, and the *hula muʻumuʻu*, a dance of a maimed person performed while kneeling, are examples of *hula* that emphasize specific body movements. The *hula ʻīlio*, dog *hula*, and the *hula puaʻa*, pig *hula*, demonstrate *hula* that honor certain characteristics of animal ancestors. Lastly, the *hula Pele* is a style of *hula* that celebrates Pele and her many family members (1998). Beamer's work introduces two more categories, the *hula aliʻi*, dance in honor of person of high political leadership status, and the *hula wahi pana*, dance for legendary or historic places (1987, 2001).

Immersive training

Being invited to participate in *hālau hula* is considered a privileged honor, though some children from *hula* families probably have little choice. Tangarō expresses it as responsible parenting: "One of the reasons children of *hula* dancers end up *hula* dancers is because everyone in the family dances *hula*, so there's no baby sitters; the only way we can be responsible for them is if they're dancing with us" (Eckardt 2013, italics added) Those *haumana* who begin immersing themselves in the *hula* traditions as children and choose to continue undoubtedly become *ʻōlapa* at an early age. Others who begin later in life need to consider putting in as much time as it would take to achieve an undergraduate degree in college, somewhere between two to five years of intensely focused study. The more time a person dedicates to immersing their lives in *hula*, the better they will become, as *haumana* are expected to master the dances they are taught—not just learn them, but master them. At the very least, this means mastering rhythmic movement, rhythmic accompaniment, and, most importantly, rhythmic attunement.

Rhythmic Movement

Every *hālau hula* assigns symbolic meaning to every movement, expression, and gesture. The same gesture in another *hālau* could have a completely different meaning, as the movements are personalized to a particular place of origin. So it is unwise to assume that similar movements mean the same thing when a different *hālau* is performing them. Even the most basic foot movement, the *kāholo*, the "vamp" step, differs between *hālau*. This step consists of a sideways movement in four counts on slightly bent knees: the right foot steps laterally a comfortable distance to the right side for the first count, then the left foot is brought alongside the right foot about hip-length apart for the second count, then this right foot, left foot move to the right is repeated for counts three and four. The *kāholo* is completed with the same movement to the left for four beats, wherein the left foot steps laterally to the left on the first count and the right foot is brought alongside the left foot for the second count and then repeated for counts three and four.

Although it appears to be a rather basic execution of foot movement, in the *ʻaihaʻa* tradition taught at Hawaiʻi Community College, the movement is done flat-footed, such that the entire foot from heel to toe is expected to be in contact with the ground for each step, whereas some *hālau* allow the heel to be slightly raised while executing this step. According to Tangarō, when the entire foot is in contact with the ground the dancer can connect with the earth, which empowers the movement. Regardless of tradition, this basic step requires dancers to maintain good balance, evenly distribute their weight, and take comfortably spaced steps, as wider steps tend to distort the fluidity of the movement.

Fluid movements are achieved with strong, flexible knees and ankles that allow the upper body to appear to float without any up-and-down bobbing motion. Comfortably bending the knees and keeping them fairly close together, so the knees are directly over the feet, which remain hip-length apart, ensures a naturally subtle hip movement. For some people this hip movement comes naturally, for others it requires hours of practice. Tangarō encouraged *haumana* to practice while washing the dishes and tending to daily personal hygiene. The more we practiced the stronger our legs and more flexible our knees and ankles would become, allowing us to execute other progressively challenging movements, especially those that require the dancer to lower the body to the ground while maintaining perfect posture.

The correct posture is probably the most important aspect of a good dancer. It begins with evenly distributed weight that is anchored in the thighs. This allows the lower torso to have the flexibility necessary to move gracefully without losing balance or timing. The strength of the upper torso is demonstrated by maintaining horizontally aligned shoulders, with the chest slightly raised at all times, especially when a movement requires the dancer to bend at the waist. There can be no shoulder slouching if the dancer is to maintain a regal elegance. The lack of shoulder movement allows emphasis on the arm movements, which are integral to the embodiment of intentionality.

The basic hand position for the *kāholo* is to extend the right arm to the right side of your body with fingers pointing to the right and palm down, while your left arm rests comfortably across your chest with elbow bent and palm down. Both hands should be relaxed, not stiff, and pointing in the same direction as the movement. So, when you move to the left, the left arm is extended to the left, with fingers pointing to the left and palm down, while your right arm rests comfortably across your chest with elbow bent and palm down. Some *hālau* require all dancers to execute this move such that their hands appear to be at the same height across the group, which means taller dancers must bend their knees lower to match the shorter dancers. In the style of *hula* taught at Hawaiʻi Community College, the hands are aligned with the level of each dancer's breasts. According to Tangarō, this acknowledges the beauty and diversity of every individual's body, encouraging each dancer to respect and appreciate the range of movement their body allows while performing the choreographed movements associated with the rhythmic accompaniments.

Rhythmic Accompaniments

While the most basic *hula* movement provides hints of the symbolic expressions maintained by each *hālau hula*, they all require some kind of rhythmic accompaniment, adding another layer of symbolic expression. Frequently used rhythmic patterns are given specific names, allowing the *kumu hula* to call out a pattern by name so the *hoʻopaʻa* can execute the associated rhythm. For example, according to Beamer, the traditional rhythm for a *mele hula* that she learned from her grandmother Sweetheart was three *kāhela* and one *pā*, where "the *kāhela* is one downbeat (u) and two upbeats (te): u te te. The pā is a single downbeat and upbeat: u te. The beginning *mele hula* rhythm pattern is thus: u te te / u te te / u te te / u te" (1987, 3, italics added).

This rhythm usually precedes any movement and signals the dancers to take their stance. For the *ʻaihaʻa* tradition, this means lowering the body by bending the knees for the *'aiha'a* stance. The rhythm used for the *kāholo* is one *kāhela* and one *pā* for each direction.

ho'opa'a (instrumentalist)	*haumana* (student)
u te te	right, left
u te	right, left tap
u te te	left, right
u te	left, right tap

The side step of the right foot for the first count is done on the downbeat, and the left foot is brought alongside on the second count, on the two upbeats of the *kāhela*. The second side step of the right foot for the third count is again done on the downbeat, but the fourth count is performed on the single upbeat of the *pā*, because the second side step of the left foot is tapped on the ground, not planted, so it can begin the first count going to the left.

After a while these named rhythms and matching movements become ingrained in the dancer's body memory; a *kumu hula* can warm up the *haumana* simply by calling out the names of the movements, and any changes in the accompanying rhythms are performed seamlessly. For example, a *kūkū,* three downbeats and two upbeats, is used on the *'ami kūkū*, a rotation of the hips with two small quick revolutions and one slower revolution. When the *kumu hula* calls out for a *kūkū,* the following rhythmic combination is performed:

ho'opa'a (instrumentalist)	*haumana* (student)
u	small, quick hip revolution
u	small, quick hip revolution
u te te	slow hip revolution

These rhythmic combinations reflect the basic lessons learned during beginning classes but do not mean these steps are executed only with these rhythmic accompaniments and vice versa. The point is to recognize that rhythmic accompaniment maintains symbolic significance. In this case certain rhythms are associated with certain *hula* movements. Specific rhythms have also been associated with specific people. My cousin once relayed to me that during an all-night vigil she heard the beating of the *pahu*, drum in the distance. She remembered the rhythm and shared it with a *kumu hula* with whom she was acquainted. She was told that rhythm was associated with a particular *ali'i wahine* and that hearing it was a good omen with regard to her vigil because it was highly likely the drum she heard originated from a different moment in space/time than the one she was occupying.

Rhythmic Attunement

The ability to reach across and relate with other dimensions of space/time is neither strange nor special for some Kanaka Hawaiʻi. It is in fact what a *hula* performance ultimately provides, an opportunity to become attuned to another moment in space/time. An *ʻōlapa* who can transport the audience into another realm has mastered not only the movements but also the art of surrendering their entire being to the *hula*. They no longer think about where their foot needs to be placed or when they need to turn because they are at one with the *hula*. At that moment, the *ʻōlapa* achieves a super sentience wherein the combination of rhythmic movement, vocal expression, and attentive intentionality merges, allowing the carefully choreographed performance to generate the vortex that connects two different points in space/time.

This is likely why, from the very beginning, some *hula* performances were considered ritualistic. Connecting to different points in space/time requires unrelenting and rigorously performed perfection. No doubt there were only certain seasons during which rituals of this magnitude could be performed, making any errors inordinately costly. Thankfully, these days, public *hula* performances do not carry the same weight of consequence, but they do allow for the committed *ʻōlapa* to symbolically connect with entities and processes larger than themselves. As Kealiʻikanakaʻoleohaililani shared in a personal communication, "The dancer is the rain, the wind, the magma." Although not every *ʻōlapa* is capable of achieving this level of mastery, nor is every observer/participant able to yield their entire being to the performance, the potential to do so still exists.

Hula implements and instruments

Musical accompaniment is an essential element in a *hula* performance, though not entirely necessary, as the oldest forms of *hula* were performed without any implements or instruments. According to Beamer, there are twenty-five implements or instruments *kumu hula* could choose to use in *hula* performances. She distinguishes an implement as a natural element, such as a piece of wood or stone, and an instrument as something fashioned by hand (2001, 2). A majority of the implements and instruments used in *hula kahiko* are percussive rather than melodic, whereas the *hula ʻauana* makes use of more modern instruments such as the *ʻukulele*, guitar, and bass. Many *hālau hula* require *haumana* to find or make their own instruments, ensuring the perpetuation of the process as well as the personalization of the product.

Emerson lists thirteen instruments used in the *hula kahiko*: four he categorizes as distinctly percussive, two as rattles, two as noisemakers, and the remaining five as wind-like instruments (1998, 140–148). Another way of categorizing these instruments/implements is according to the divine entity(ies) from whose *kinolau* they consist. For example, the most important instrument according to Emerson is the *pahu*, because its deep and solemn tones make it "an instrument to stir the heart to more vigorous pulsations, and in all ages it has been relied upon as a means of inspiring emotions of mystery, awe, terror, sublimity, or martial enthusiasm" (1998, 141). Adrienne Kaeppler and colleagues say it owes its importance to its association with those *hula* performances conducted during temple rituals and that it lost its distinctiveness when it was incorporated into mainstream public performances (1993, 1–2). Perhaps its importance has more to do with the fact that the base, a hollowed coconut trunk, is made from the *kinolau* of Kū, and the membrane, sharkskin, is recognized as *ʻaumakua*, the trusted divine entities who assist and guide Kanaka Hawaiʻi. Together this balanced pair of land and ocean elements resounds when struck with the palms or fingers, sending sound waves both into the earth and bouncing back up into the atmosphere. Classifying instruments as such gives greater symbolic relevance to the performances.

For example, when the Hawaiʻi Community College beginning *hula* class learned a *hula* using *ʻiliʻili*, small smooth flat stones, Tangarō explained the symbolic connection between the implement and the *hula*. *ʻIliʻili* originate from the bowels of the earth, beginning as magma that went through a birthing process of sorts and was eventually coaxed down from the mountains along waterways that also shaped them into the dense, smooth, flat stones. They are the products of Pele and Kāne, and striking them together allows them to speak and represent those divine entities and processes. He made it abundantly clear that, since each dancer needs to personalize their implements and form their own relationships with the divine entities and processes that they represent, each *haumana* had to search for their own *ʻiliʻili*. It made them all the more precious. Furthermore, if and when the purpose of taking the *ʻiliʻili* ever came to an end, as when the *haumana* no longer chooses to *hula*, then the *ʻiliʻili* needed to be returned to the place where they were found. Intentionality matters.

ʻAʻahu and adornments

Dancers are also expected to make their own *ʻaʻahu*, apparel, and adornments, as most people tend to take better care of anything they have had to invest time in creating. This begins with making the *pāʻū*, skirt, used for practice sessions. While most *hālau hula* instruct *haumana* to use an off-white muslin fabric folded lengthwise and gathered into pleats with a cotton rope cord inserted along the length of the fold so it can be tied at the hip, securing the skirt in place, a resurgence of *kapa*, bark cloth, and handwoven cordage is currently taking place. In 2011, Hālau o Kekuhi took the stage at Merrie Monarch donned in *kapa*, while over thirty *kapa* makers sat in the front row for the performance.

This collaboration between *hula* and *kapa* practitioners was sixteen months in the making. Reconnecting two cultural practices that remained mostly separate arts throughout the period of the 1970s commonly referred to as the "Hawaiian renaissance" was a huge undertaking, as over 350 square feet of *wauke*, paper bark, was stripped, soaked, pounded, and dyed for the eighteen performers and three *kumu hula* that took the stage. According to 2015 Native Arts and Cultures Foundation Traditional Arts Fellow Dalani Tanahy (quoted in Reiny), the 2011 Merrie Monarch performance "could very well be the first *hālau* to perform with *kapa* since Kaʻahumanu's time" (Reiny 2011, italics added).

According to Tanahy, it all starts with good trees. *Wauke* grows quickly and can reach six to twelve feet in height in a couple of years, when it is best for harvesting. The harvested trees need to be stripped and soaked within a few days, or the material will be difficult to work with. The initial pounding is done with a *hohoa*, round beater, and a *kua pōhaku*, stone anvil, to loosen the fibers, coaxing it into an even, pulpy consistency. It is coiled, wrapped, and stored in water for a couple of weeks to soften. The second beating is done with an *ʻiʻe kuku*, square beater, on a *kua laʻau*, wooden anvil, to gently compress the fibers while flattening the material to the desired width. The watermarks are added in the final stages of felting using the engraved side of the *ʻiʻe kuku*. The *kapa* is then laid out to dry but needs to be rubbed with stones for a smoother and softer texture before dyes can be applied. Dyes are made from various parts of many different plants, and the *kapa* is immersed in the colorful liquids until the color is set. *ʻOhe kāpala* (bamboo stamp) designs are dipped into other dyes and pressed onto the *kapa* in repeating patterns. Some *kapa* were even scented (Tanahy 2006). As you can see, the processes could very well be cartographic in their own right, as each step undoubtedly incorporates symbolic significance, from the making of the tools and dyes to the watermark pattern impressed and the design imprinted.

Oftentimes *ʻaʻahu* are made and worn for specific performances and are carefully cared for, covered, and closeted when not in use. Several protocols are associated with the use and handling of *ʻaʻahu*, including the one used for practice at the *hālau*. At the Hawaiʻi Community College beginning class, Tangarō instructed us on how to make the practice *pāʻū*, but required us to bring the final product to him for inspection before we could use it in the *hālau*. Upon approving the materials and construction, he then for the first time tied each one on the student who made it, signifying another rite of passage similar to the *oli kāhea*.

Other adornments are carefully made and fastened on the head, shoulders, wrists, and ankles, depending on the type of *hula* being performed. Even the simplest adornment of woven ferns requires specific protocols, from gathering and designing to securing them in place for a performance. Most materials for the *hula kahiko* are found in the *wao akua* and require protocols similar to those expressed earlier in the dressing of the *kuahu*. Dressing the dancer is no different. Both Emerson and Beamer share that the act of securing each article of the regalia in place was accompanied by *mele* (Emerson 1998, 49; Beamer 2001, 6–7). According to Beamer,

> At a given point in the chant, we began to lean forward to pick up our greenery, to begin the dressing ceremony, the ferns for our wrists would be put on first. Each dancer would lay the right wristlet over her right knee, place her right hand on it, palm up, and with the left hand, twist the ends of the wristlet and tuck it securely under the band. When that was complete, we held the wrist to admire our handiwork. Then both hands were place in a relaxed pose on the lap, palms up. The same procedure was followed for the left hand. Next came the anklets, beginning with the extension of the right foot diagonally

> forward, holding the anklet up as though admiring its beauty, bending to the right side, fastening the anklet at the back of the right ankle, returning the right foot to place and sitting back. At this point, there was a deep breath as we cupped our hands in our laps in a humble attitude. The same procedure then began for the left ankle. (2001, 6)

She continues by describing the placement and securing of the *lei poʻo*, head garland, and the *pāʻū*, culminating with the *lei ʻāʻī*, neck garland. When all adornments were placed, a final *mele* was performed, signifying that the dancers were fully aware of their commitment and were ready to sacrifice their mind/body to become the dance, entering a near trance-like state of being.

Regardless of the venue, when it is time to perform, *ʻolapa* stand ready. They are silent, focused, and completely confident in their preparation. At this point, they are no longer individuals; they are the medium through which the divine entities and processes of the natural world are expressed, bringing to life the space/time the *mele* describes. The *kumu hula* begins to *oli kāhea*, signaling to the dancers and the observer/participants that the presentation is about to begin. As the *kumu hula* and the *hoʻopaʻa* prepare the ground for their instruments, the dancers arrange themselves for their entry. Their look is solemn, their dress impeccable, their breathing metered, and their presence undeniable. What happens next depends on the type of *hula* being performed. The next section describes one of the simplest *hula* I learned as a student at Hawaiʻi Community College's beginning *hula* class.

Ka Hōʻike (The Performance)

"ʻAi Kamumu Keke"

ʻAi kamumu keke
Crunching cinder consumed by fire
Nā keke pāhoehoe ke
Smooth unbroken lava rendered to cinder
Wela i luna o
The fire rises of
Halemaʻumaʻu ke
Halemaʻumaʻu

'Ai kamumu keke is a *mele* describing the sight, sounds, and essence of magma rising in Halema'uma'u, the crater in Kīlauea caldera considered the metaphoric home of Pele. It is a timeless composition, as the event that inspired it continues to this day. When the *kumu hula* choreographed this *mele*, particular attention was paid to the characteristics of lava, also known as *pele*.[30] When lava first emerges from the depths of the earth, it is liquid fire—sometimes thick and slow-moving, creating *pāhoehoe*, smooth lava, and sometimes highly gaseous and quick-moving, creating *'a'ā*, rough lava. As the *pele* rises in Halema'uma'u crater and consumes the semi-cooled basalt crust at its surface, it crackles, pops, and crunches, making indistinct rumblings of something approaching. To find the source of this particular kind of noise, one need only look toward the skies above Halema'uma'u, lit red by the rising magma.

The dance style the choreographer chose to use is a specific type of *hula ki'i*. *Ki'i* is an image, statue, idol, or doll. A *hula ki'i* can take the form of dancers either using marionettes or positioning their bodies like a carved idol. In this case, the choreographer uses the latter type of *hula ki'i*. In this *hula*, the dancers can be standing or sitting, as their feet do not move. The arm movements are exactly the same, regardless whether the dancer is standing or sitting. One of the main reasons this *hula* was selected for this description is that each of the four lines has one main movement, making it easier to describe without taking away from the Kanaka Hawai'i cartographic presentation being expressed in this section. (A snapshot of the first three arm movements are included on this book's cover.)

The "ready" pose for this *hula* looks similar to the stance some people take when they are upset or angry. It begins with the dancer either sitting on the heels, similar to a yoga "hero" pose, or standing with the heels together, toes pointed at a forty-five-degree angle, with knees slightly bent in same direction as the feet. The arms are at the sides, with elbows bent outward away from the body at a forty-five-degree angle, and the palms facing the floor with the wrists bent at a forty-five-degree angle. The fingers are pointing outward, mimicking the direction of the toes of the standing pose. The angularness of the pose also mimics some Hawai'i temple woodcarvings. I believe the motionless lower body puts more emphasis on the arm movements, represents the earth as a steadfast foundation, and gives the dancer a constant connection with a solid foundation, making it easier to commit to the natural processes the dance describes.

The first arm movement looks as though the dancer is playing peek-a-boo, covering the face with the hands. It begins with the arms out to the sides, parallel to the ground at shoulder height, with elbows bent such that both hands cover the front of the face without touching the face. Both palms are turned toward the face, with one hand in front of the other and the fingers and thumbs bent like claws. Every muscle in the dancer's body is taut and stiff as the hands slowly move outward to the sides, exposing a face wearing a fierce expression. The movement conveys the powerful, steady, expansive movement of the *pele* as it rises in the crater, breaking up the semi-cooled basalt crust as though fingers were scratching at the surface, seeking freedom from the fiery depths. The words that accompany this movement describe the sounds as crunching cinder, and further research of the words reveals it is also a thunderous crackling rumble or the confused noise of a multitude at a distance. The accompanying *oli* is forcefully loud and booming.

The second arm movement looks as though the dancer is hoisting a large heavy bowl up above the head. The dancer keeps the arms at shoulder height, parallel to the ground, but bends the elbows upward, making the forearms nearly perpendicular to the ground. The wrists are bent with palms facing up to the sky and the fingers are still clawlike. The dancer's head is tilted slightly upward, with the eyes looking toward the sky as the arms are raised upward. The movement represents the *pele* rising in Halemaʻumaʻu crater as the accompanying words the dancer chants express how this movement affects the *pāhoehoe* crust on the crater floor, being rendered to cinder.

The third arm movement looks like the dancer is shielding the forehead from the sun. The dancer maintains the arms at shoulder height with elbows bent outward to the sides and head tilted slightly upward. The hands are brought together near the forehead with the palms facing upward. The fingers are straight and the fingertips are nearly touching. The words that accompany this movement represent the magma rising, which can be seen from a distance through the reflection on the red-tinged clouds that hover over Halemaʻumaʻu.

The last movement begins with the dancer looking like a scarecrow and ends with the dancer making a large bowl with the arms. For the first part, the dancer maintains the arms at shoulder height but bends the elbows downward, making the forearms nearly perpendicular to the ground with the palms facing the back. For the second part, the hands move forward in an arced path, making a circle in front of the body with the palms facing outward and fingers touching slightly. The movement represents Halemaʻumaʻu crater, the source of the rumbling noise. In the first part of the movement, the dancer uses the body and arm positions to create the image of a *hale*, a house. The symbolic significance of this movement highlights the crater as the home of Pele, from where Kanaka Hawaiʻi are able to witness the event this *mele* describes.

According to Kekuhi Kealiʻikanakaʻoleohaililani, this *hula* transmits important knowledge to the observer/participant who is unfamiliar with who Pele is and what Pele does. The dancer demonstratively expresses the power generated as Pele erupts, using stiff, tight, purposeful body movements and a loud, explosive chanting style. The face is fierce, the teeth are bared, the muscles are taut and stiff like a rock generating a lot of heat in the dancer's body, making it possible for the dancer to embody the power of Pele, not just dance about Pele. Kealiʻikanakaʻoleohaililani further states, "As long as the lava still flows and the sun continues to remind us of the heat of Pele, the stories are never relegated to the past—they are now. When the vog comes, Pele is here now. Pele is not an entity with nose, eyes, mouth, arms but she is lava, rock, sulphur, vog, red lava, lightning that come from the lava, all of that smoke" (2015).

Many Kanaka Hawaiʻi do not distinguish between the natural entity and processes known as *pele*, and the divine entity Pele. According to Kanaka Hawaiʻi cartographic philosophy, presented in part 1 of this book, the natural entity and the divine entity are one and the same. Categorizing natural entities and processes according to their physiological characteristics is not the same as anthropomorphizing the environment, because Kanaka Hawaiʻi do not assign human characteristics to them. In a Kanaka Hawaiʻi cartographic philosophy, natural entities and processes are recognized as living beings having an innate intelligence and roles and responsibilities. They are classified according to the space/time each occupies and the life-giving purpose each fulfills.

This chapter is not meant to present *hula* practice in its entirety, but only to highlight the cartographic processes of spatial/temporal knowledge transmission. The *hula Pele* example shared in this section both maintains and communicates spatial/temporal knowledge about Pele. Oftentimes people see only the destructive nature of Pele, inspiring fear as it exposes the frailty of humankind in relation to Pele. This is why *kumu hula* add a transitional piece that softens the performance, signifying that everything is calm as the forest can begin to grow anew. This inspires the observer/participant to step back from the human body scale of the event this *hula Pele* presents and understand it from an island scale, as one of Pele's responsibilities involves the birthing of new lands, providing more space for life to continue to flourish. Pele also inspires flora and fauna to evolve, as constant eruptions allow various species to adjust to environmental shifts including climate change and shifting planetary positions.

Perceptions of Kanaka Hawai'i Spatial/Temporal Knowledge Transmission

Hula performances are the perfect multisensual delivery system of Kanaka Hawai'i spatial/temporal knowledges. Every *hula* performance stimulates the aural, visual, and olfactory senses. Aurally, the *hula* is accompanied by both instrumentation and song, adding even more depth to the presentation of chanting styles. Visually, the bodily movements, attire, and adornments work together to form a cohesive symbolic representation of contextual essence. Depending on the type of adornment, performances can also awaken the sense of smell. Fragrant *lei* moving with the dancer emanate aromas on gentle breezes and enhance the visual and aural performance, allowing the audience to be surrounded in the scents of nature. Most people know how a certain smell can bring forth fond memories of moments long past. The same is true with wafting aromas of floral adornments. Every *hula* movement transmits a specific meaning, with gestures that symbolize flowers and animals, and natural elements like wind, water, cliffs, trees, and even conflict and war, allowing dancers to evoke an endless array of significance with expressive hand gestures and undulating body movements.

When presented with the precision and accuracy of its moment of origin, the *hula* becomes "a vehicle, very capable of pitching you into another world, into that event for which the *mele* was composed. That is what dancing *hula* is capable of doing. When that happens, and you look at the dancer, the dancer becomes the dance" (Kanahele 2005, 26, italics added). A dancer embodies an event occurring in a space/time deemed worthy of timelessness, transmitting spatial/temporal knowledge through symbolically choreographed movements, rhythmic and vocal accompaniments, and appropriate adornments. The observer/participant can become disembodied from the space/time of the performance and mentally transported into the space/time of the event being memorialized through a mesmerizing, trance-inducing engagement of multisensual arousal, regardless of whether or not firsthand spatial/temporal knowledge of the place or event being presented has been experienced. However, the more intimate the observer/participant is with the place or event being presented, the deeper the sensual experiences are capable of being.

Hula is also key in maintaining spatial/temporal knowledge for the community it serves. "Dancing is a tool to educate ourselves, telling us who we are" (Kanahele 2005, 26). *Hālau hula* maintain and transmit vast bodies of knowledge using a variety of protocols, teaching techniques, and mnemonic devices to create symbolic connections between sound, smell, sight, and movement. In this world of high-tech information transfer, data exchange, and hypermedia, many of us have become numb to other modes of knowledge transmission, modes such as *hula* that use symbolically embedded performance as a means to move local knowledge beyond the space/time of its production. Performance-based knowledge transmission is successful in *hālau* because of the methods used to maintain a specific compendium of spatial/temporal knowledge. Regardless of whether or not the dancer becomes a *kumu hula*, spatial/temporal knowledge transmission is demonstratively passed from generation to generation, infused into the moment the dancer embodies the dance with rhythmic movements, mesmerizing beats, and wafting fragrances that unfurl a metaphoric logic specific to each *hālau*.

According to sociologist David Turnbull, transmission of knowledge in oral societies is "heavily dependent on metaphors, narratives, redundancy, concrete models and communal interaction" (2003, 152). Furthermore, incorporating spatial knowledge in songs and rituals, learning and testing knowledge retention in small groups using mnemonics, and having overlapping methods that reinforce each other are effective ways to ensure that the vast body of information is not only accurately retained and passed on to future generations but also instantly accessible (Turnbull 2003, 153).

Hālau spatial/temporal knowledge transmission involves mastery of *mele*, *oli*, movements, musicality, and the making of implements, instruments, and adornments through an intensified and stringent mentoring process. *Hālau* exude formality in ceremony, ritual, etiquette, and protocol at every step of the learning process. From the moment a *haumana* decides to step into a *hālau* for the first time to the moment they step onstage to perform, and even after the performance, when they must return their *lei* to a natural setting, everything is done with a deep respect for the practice. *Hula* provides the vehicle for many Kanaka Hawaiʻi to recognize and deepen their spatial/temporal relationships with all natural elements and processes, thereby arousing their sensual engagements to a heightened state of consciousness.

CHAPTER SEVEN

Ka ʻŌlelo Hope

(Conclusion)

Kanaka Hawaiʻi cartography is a product of a Kanaka Hawaiʻi cartographic philosophy that recognizes the life essences and intelligences of all entities and processes; that nurtures intimate connections with these entities and processes, relating to them as familial beings according to standard forms of etiquette; and that cultivates metaphoric associations with each being, classifying them according to their characteristics, roles, and responsibilities. By the time Europeans arrived on the shores of Hawaiʻi, Kanaka Hawaiʻi cartographic practices reflected the philosophical and technological advancement of Kanaka Hawaiʻi; those sensual, intimate, and multidimensional relationships Kanaka Hawaiʻi nurtured with divine entities and processes are still inextricably intertwined in Kanaka Hawaiʻi cartographic performances. Kanaka Hawaiʻi cartographic practices need not share the same kind of materiality and apparent reproducibility as Western cartographic practices to serve as an effective means of pursuing representational solutions for relational and spatial issues.

In the time it has taken to write this book, a small number of scholars have begun reframing "the ontological foundations of cartography, moving from a representational to a processual understanding of maps, from ontology (what things are) to ontogenetic (how things become)" (Kitchin et al. 2013, 480). Kanaka Hawaiʻi cartography builds on the works of these scholars, further extending the reconceptualization of the ontogenetic positionality of cartographic practice by challenging the apparent philosophical stability cartographic practices have been required to maintain and providing the alternative philosophical underpinning, presented in part 1, that laid the foundation for understanding how Kanaka Hawaiʻi cartographic consciousness is framed.

The Kanaka Hawaiʻi philosophical perception of the world described in this text is based on an ontological understanding that everything in this world, including natural entities and processes, has a life force and, as such, all are considered intelligent beings with roles and responsibilities. Many Kanaka Hawaiʻi also recognize these distinctive intelligences by using an epistemological framework based on kinship and interrelatedness, such that natural entities and processes are our primary teachers, with whom dynamic relationships are commonplace. To provide a structured process of learning, Kanaka Hawaiʻi methodologically employ both *kinolau* and *kaona*, grouping characteristic likenesses and expressing pluralities of meaning, respectively. *Kumu hula*, Hawaiʻi dance teacher, and Hawaiʻi scholar Pualani Kanakaʻole Kanahele of the Edith Kanakaʻole Foundation uncovered formal modes of knowledge organization, *papahulihonua, papahulilani*, and *papanuihānaumoku*, in the Kumulipo, a *koʻihonua*, cosmogonic genealogical chant, that recites birth order of many natural entities and processes. She considers these the foundational layers for understanding all life and existence in the Kanaka Hawaiʻi reality.

All human beings have similar sensory receptors, with about the same physiological capabilities with which to receive information about the world. People in different places throughout time mix the information received by their sensory receptors, both physically attained and culturally attuned, in varying proportions, placing greater emphasis on different sensory systems. These systems provide us with information about the world around us. They mediate our experiences both through their structures and through the way we use each system. Our sensuous experience and perception of the world "is grounded on previous experience and expectation, each dependent on sensual and sensory capacities and educational training and cultural conditioning" (Rodaway 1994, 5). Both Meyer and Oliveira acknowledge that Kanaka Hawaiʻi acquired spatial knowledge empirically through their body's sensory systems. However, Meyer emphasizes the cultural variability of these sensory systems, and Oliveira extends the five commonly recognized sensory systems with four additional genealogically embodied sensory systems: *naʻau*, ancestral lineage; *kulāiwi*, place lineage; *au ʻāpaʻapaʻa*, time/place/space alignment; and *moʻo*, succession of ancestral responsibility, all of which are considered something a child is born with or born into (Oliveira 2014).

These philosophical underpinnings inform two important cartographic concepts: spatial/temporal orientation and classification. The integrated abstraction of space/time is a Kanaka Hawaiʻi cartographic concept, derived from the terms *wā* and *manawa,* which are explained both as a space or interval between two objects or events and as a period of time. This adds another level of complexity to the "spatial turn," or the shift in focus from "the theoretical privileging of time over space in discourses of modernity . . . [toward] . . . a critical re-envisioning of the *spatial* . . . for which 'the map' has become the defining trope" (Roberts 2012, 14, italics original; see also Foucault 1980; Soja 1989). The integrated concept of space/time shifts focus away from the "things" each represents and toward the "engagements" the combined entity generates.

To facilitate discussion of Kanaka Hawai'i cartographic perspectives on the key concepts of orientation and classification, a Kanaka Hawai'i understanding of scale was introduced based on relative sizes of key entities, beginning with the human body, then the island, then the planet. Presenting the Kanaka Hawai'i understanding of these concepts in this fashion does not mean they are independent from one another. In fact, engagements at the planetary scale directly affect the body scale and vice versa. Through the body, Kanaka Hawai'i understand relative direction, as different entities move relative to one another. At the island scale, the concept of absolute direction is observed based on the movement of celestial entities against the backdrop of an island that appears to be stable. The planet scale provides an understanding of the concept of cyclical and seasonal space/time engagements.

Orientation of the body, island, and planet is a fairly straightforward physiological observation. Although it may seem odd to use the term *physiological* when speaking of an island or a planet, it is a natural outgrowth of recognizing all entities. Keep in mind this is not the same as anthropomorphizing nature. That point of view cannot exist in a kin-centric consciousness. Kanaka Hawai'i do not relate to more-than-human entities as though they were human. They relate to each of the thousands of land, ocean, atmospheric, and celestial more-than-human entities as though they each have their own innate intelligence, with specific roles and responsibilities. A kin-centric consciousness allows Kanaka Hawai'i to learn from these more-than-human entities, classifying each according to their characteristics, roles, and responsibilities.

Classification at each of these scales brings in a sociopolitical element. The body is not only biologically classified according to age, size, or gender; classification also includes social and professional status. Things get a bit more complicated at the island scale. Kanaka Hawaiʻi did not see just physiological classification at the island scale; they also established socio-political classifications to maintain sustainable resilient relationships with the places where they lived. Lastly, classification at the planet scale led to the identification of divine entities and processes. The Kanaka Hawaiʻi cartographic foundations and framework presented in part 1 determine the kinds of knowledge Kanaka Hawaiʻi believe they can acquire, shapes the way that knowledge is symbolized or represented, and affects the processes used to communicate or transmit that knowledge to others.

Kanaka Hawaiʻi acquire spatial/temporal knowledge empirically, sensually, through cultural lenses that are simultaneously objectively subjective and subjectively objective because of the value placed on the balance between the metaphoric and the rational mind/body. The "metaphoric mind" is a distinction Tewa educational scholar Gregory Cajete describes as encompassing the perceptual, creative, and imaginative experiences that develop from birth until language is learned, at which time the rational mind develops and is integrated into the metaphoric experiences (2000). Kanaka Hawaiʻi practitioners respect each for their unique abilities, as neither is consistently overemphasized when intimately engaging with the world or understanding the roles and responsibilities of all beings in the world.

In many societies, the rational mind becomes more prominent when language becomes literacy. The transition from an oral culture to a literate culture is a shift in emphasis, from practices that value a multiplicity of meaning to those that prefer the stability of standardized singular truths. Writing has a tendency to fix any account being transmitted, removing the adaptability and innovation of metaphoric multiplicity. When the memories of a culture begin to be transmitted mainly by the reproduction of their inscriptions, rather than by "live" telling, improvisation becomes increasingly difficult and innovation is institutionalized, capitalized, and commercialized.

Thankfully the Hawaiʻi language persuasively conditions the mind/body to intimately and sensually participate with natural entities and processes. For example, the term for meal, *ʻaina*, literally comes from the term for food, *ʻai,* and it is mischievously similar to the terms for copulation, *aina*, and land, *ʻāina*. These differences are easy to distinguish in written form but not necessarily so when spoken, unless you happen to be a Hawaiʻi language speaker. But wait, there's more: *ʻainā* is described as sore aching, stiffness, as from overexercise, and *ʻaʻina* is a crackling, snapping, sharp report that can be either an explosive or a long prolonged sound (Pukui and Elbert 2003). Now, my impish character automatically goes to the gutter with the explanations for these terms, but what amazes me is how these closely related sounds appear to provide closely related contexts.

Thus, because of the metaphoric flourish embedded in the Hawaiʻi language, Kanaka Hawaiʻi scholars well-versed in Hawaiʻi language are able to recalibrate our sensual engagement as they conduct archival research of documents written in the Hawaiʻi language. A great example is Kanaka Hawaiʻi historian and archivist Noelani Arista, whose archival research techniques enable her to recontextualize documents written decades ago using a distinctly Kanaka Hawaiʻi framework. This is no easy task by any measure, because she also maintains that the ideas generated must retain the perspectives of the time period from which they are derived.

Kanaka Hawaiʻi spatial/temporal knowledge representation reflects metaphoric modes of classification and expression. While metaphoric relationships are fundamental parts of human cognition, helping to make unfamiliar concepts more understandable in terms of familiar ones (Mark 1993), each culture creates and maintains their own metaphoric relationships based on their understanding of reality. Two specific methodological devices of Kanaka Hawaiʻi cartographic significance are the use of *kinolau* and *kaona*. Both indicate a propensity for a multiplicity of meaning that is ever present and embedded in the Hawaiʻi islandscape. Divine entities and processes are named according to embodied characteristics, roles, and responsibilities. For example, freshwater is considered the realm of the divine entity Kāne; however, rains that are heavy and penetrating, such as those experienced in the winter months, are attributed to the divine entity Lono; and when freshwater takes the form of light mists that provocatively cling to the upper reaches of the mountain foliage, tenderly nurturing them with trickles of water, it is the divine entity Laka. While *kinolau* are metaphorical associations of divine entities and processes, *kaona* is an artistic or poetic composition imbued with symbolic meaning. For example, rain is used in *mele* to describe both a tantalizing sensual escapade as well as the near-debilitating loss of a loved one. It depends on context.

Kanaka Hawaiʻi spatial/temporal knowledge transmission connects the observer/participant to the space/place of its origin through rigorously maintained mentoring. In all Kanaka Hawaiʻi cartographic performances described in this book, practitioners began their training through direct observation and listening. Survival meant knowing how to determine the location of a food supply, like fish in the ocean. Being able to see the movement of fish in the water or correlate the circling of birds with schools of fish is not a skill everyone possesses, but those that do, in many Kanaka Hawaiʻi households, are capable of becoming great fishermen/women. In other places, survival also meant associating the sounds of rocks exploding underwater in specific locations such as Kepuhi with oncoming storms that could threaten the lives of nearshore occupants.

Kanaka Hawaiʻi mentors require successors to have excellent memory capabilities and utilize metaphorically imbued compositions, rhythmic timing, and symbolically embodied movements as mnemonic devices, transforming essentially unstructured series of phenomena or information into seemingly coherent compositions. Perfect recitation of information is necessary for the maintenance of particular traditions. *Haumana* practice until they can confidently recite all information shared without error, as, at one time, mistakes made during ritualistic performances resulted in death. Although slipups are not usually met with loss of life these days, they still carry consequences.

Additionally, Kanaka Hawaiʻi mentors do not usually entertain numerous questions. They expect *haumana* to reflect on their inquiries themselves before raising their hands or jumping to conclusions. This forces each apprentice to find their own way through the lessons, forming personalized solutions rather than having the answers given to them. In this way the training process encourages critical thinking and responsive actions. Instead of asking *haumana* if they have any questions, mentors engage them experientially, giving them opportunities to demonstrate their learning while mistakes can still be corrected. Only after observing, listening, reflecting, and doing are questions entertained. By this time, the only questions that remain are those that are the most important for the learning process. This process ensures independent learning through careful guidance while mastering particular skills and provides *haumana* the opportunity to participate in long-held traditions of knowledge maintenance. This form of training is not just about passing on specific skills or sets of knowledge. It is about preparing the *haumana* to engage with the world systematically and encourages a manner of dealing with any crisis or situation that occurs that was not part of the training process. For example, when Thompson was sailing blind, moving swiftly across the ocean without a solid sense of direction, holding the lives of his crew and the hopes and aspirations of his family and community members in the balance, his initial dread came from his reaction to failure; his success came from the processes he had encountered under the careful guidance of his mentor, Mau.

Kanaka Hawaiʻi cartography did not grow from the seed of measured precision. It grew from the seed of revered relationships. Both are rigorous. Both are regimented. Both require the processes of knowledge acquisition, representation, and transmission to achieve their respective resultant "mappings," those spatial/temporal forms "brought into being and made to do work in the world (e.g., inscribing territory, shaping discourse, producing knowledge, informing and framing decision making)" (Kitchin et al. 2013, 481). Kanaka Hawaiʻi maintain several cartographic practices that provide contextually specific spatial/temporal solutions, such as *ka hoʻokele, ka haku ʻana*, and *ka hula*.

Each specific practice presented in this book highlights one of the three basic cartographic processes—spatial/temporal knowledge acquisition, representation, or transmission. *Ka hoʻokele* depicted sensual awakening as a necessary part of the spatial/temporal knowledge acquisition used in Kanaka Hawaiʻi cartographic performances. *Ka haku ʻana* provided ample descriptions of the sensual awareness achieved through metaphoric and symbolic representations of spatial/temporal knowledge. *Ka hula* revealed how attentiveness to traditionally maintained spatial/temporal knowledge transmission processes allows rhythmically sensual movements to render distant places or times accessible. Though presented as separate practices, each engages in all the basic cartographic processes. Careful attention was taken to ensure an understanding of the preparation necessary for each cartographic performance. Becoming a well-respected *hoʻokele, mea haku, kumu hula* requires years of sacrifice and discipline. It is no less important than the educational corporation now dominating American society. In fact, it could be argued that these kinds of mentoring processes can lead to a more caring and productive populace.

Kanaka Hawaiʻi cartographic practices and their associated mappings continue to evolve, while maintaining an intimate and embodied engagement with nature. I believe *hula* is the epitome of Kanaka Hawaiʻi cartography because its performances "unfold in context through a mix of creative, reflexive, playful, affective and habitual practices; affected by the knowledge, experience and skill of the individual to perform mappings and apply them in the world" (Kitchin et al. 2013, 481). Much like a map is an embodiment of a Western conception of measured space, a *hula* performance is an embodiment of a Kanaka Hawaiʻi perception of an intimate interaction with the natural world. Such performance presents compact concise situated knowledge and sensual interactions about past events and honored ancestors, both human and more-than-human. It is an expression of what is sensually experienced, put to symbolic movement and elevated by the rhythmic sounds of musical accompaniments and wafting scents of floral adornments. It is an integrative presentation of experienced space providing a deeper layer of knowing and relating to the world.

Eia Au

(I Am Here)

Now that I am at the end of this journey I can say with complete certainty that I have been changed as a result of it. When I began this book, my only intention was to make a worthwhile and stimulating text for academic audiences that would still be accessible for the general reader. Although I've learned that a career in academia might not be the best fit for everyone, I still believe we need more Indigenous voices and perspectives in academia to shift the collective intelligence toward research based on moral relationships with our more-than-human associations. And while I have referenced as many Kanaka Hawaiʻi scholars as possible, I still did not touch on the important works of people such as *kumu hula* Kalani Akana and his research on *hei*, Hawaiʻi string games, or noted cultural practitioner Keone Nunes and his role in the revival of *kākau*, Hawaiʻi tattooing.

I am so very grateful to those who have maintained Hawaiʻi practices and the knowledge they preserve. In my opinion, it is because of these people that the Hawaiʻi renaissance of the mid-twentieth century successfully motivated many Kanaka Hawaiʻi to attain degrees in higher education, bringing the Kanaka Hawaiʻi voice forward and, in the process, reinvigorating Hawaiʻi cultural practices. Generations of Kanaka Hawaiʻi academics and practitioners fought to create a space for Kanaka Hawaiʻi intelligence and ingenuity, and now there are several Kanaka Hawaiʻi scholars and practitioners sharing the knowledge of our ancestors in various disciplines, including but not limited to history, literature, political science, geography, anthropology, education, biology, and marine biology. I sincerely appreciate the works of each of the scholars I have referenced and recognize their efforts to extend the margins of academic thought. Their struggle to be heard in their respective disciplines is a humble reminder that we are all working toward a similar goal—bringing the Indigenous voice out of the shadows of misrepresentation and toward enlightened awareness.

It is my hope that describing Kanaka Hawaiʻi cartography encourages other Indigenous peoples to recognize and embrace their own cartographic traditions so they can expand the current understandings and representations of the world. Our knowledge need no longer be subsumed or assimilated into Western knowledge systems, but can stand side by side with other knowledge systems as a viable expression of spatial/temporal engagements.

There is so much more to be done. My presentation on Kanaka Hawaiʻi cartographic practices is not exhaustive, nor am I a master of any of these practices. Fortunately, neither is necessary to adequately and accurately identify the cartographic elements embedded within the practices. I engaged in this undertaking to establish a foothold for the next generation, who will be more than capable of extending this work as fluent Hawaiʻi language speakers and cultural practitioners. The wisdom of our ancestors continues to provide the knowledge we need to survive in our homelands. It is our ongoing responsibility to ensure that the information we share does not become merely data in someone else's knowledge framework—that wisdom and knowledge are not reduced to information and data. I look forward to the accomplishments of the coming generations.

In parting, I share my gratitude with all who have shared their wisdom so another generation may flourish, those who continue to maintain our Kanaka Hawaiʻi knowledge traditions, and those who have accompanied me on this journey.

"Oli Mahalo"

Ūhola ʻia ka makaloa lā
Unfurled is the makaloa mat
Pūʻai i ke aloha ā
Food shared in love
Kūkaʻi ʻia ka hā loa lā
Exchanged is the long breath
Pāwehi mai nā lehua
The lehua honors and adorns
Mai ka hoʻokuʻi a ka hālāwai lā
From zenith to horizon

Mahalo e Nā Akua
Gratitude and venerated admiration to the multiplicity of divine entities
Mahalo e nā kūpuna lā, ʻeā
Gratitude and appreciation to our beloved ancestors
Mahalo me ke aloha lā Mahalo me ke aloha lā

Gratitude, thanks, and love (to all beings present, both seen and unseen).

Glossary

ʻā- —In the nature of the noun that follows

ʻaʻahu—Clothing

aʻe—Upward, on top of, above, on

ʻaekai—Seawater edge

ʻaha—Coconut braided sennit cord

aheahe—Hawaiʻi vocal quality of loud, forceful articulations contrasted with soft, gentle sounds; see *haʻanoʻu*

ahi—Fire, red

ahiahi—Evening

ahupuaʻa—Land area; a land division within a *moku* named because of the stone altar erected along the path as one enters the land section where residents would place offerings as part of the *makahiki* celebration

ʻai—To eat or take a bite; food, specifically derived from nonmeat sources; metaphorically, to rule

aia i hea au? —Where am I?

ʻaihaʻa—Hula style danced with bended knees

ʻaikapu—Era of restricted eating

ʻāina—Land, specifically that land from which a person eats; Hawaiian Star Compass house where islands are found, as Hawaiʻi is at 21 degrees north and Tahiti is at 18 degrees south

aka—Shade or shadow

ʻākau—Right side, in the nature of the summer season

ākea—Broad, wide, spacious, open, unobstructed

aku—Away from the body

akua—Divine entity associated with life-giving natural process(es)

alakaʻi—Leader

aliʻi—Hawaiʻi political leader

aliʻi ʻai ahupuaʻa—Hawaiʻi political leader that eats from the *ahupuaʻa*

ali‘i nui ‘ai moku—Hawai‘i political leader that eats from the district
ali‘i wahine—Hawai‘i female political leader
alo—Front
aloha—Compassion, sympathy, kindness, grace
alolua—Opposites
‘anae—Mullet fish
anahulu—Ten-day period similar to a week
‘ane‘i mai—On the near side of the body
ao—Light, day
a‘o—Learning, teaching
a‘o aku—Teaching
a‘o mai—Learning
‘āpa‘a—Dry area with small trees
‘Āpa‘apa‘a—Strong thrashing winds associated with Kohala
au—Period of time, current, flow, movement, space/time
Au ‘āpa‘apa‘a—Oliveira's sense of place/time; specifically, how different places affect our perception of time and vice versa
‘auina—Descend
‘auinalā—Late afternoon
‘auinapō—After midnight
‘aumakua—Trusted divine entities who assist and guide Kanaka Hawai‘i
aumoe—Midnight
awakea—Late morning to early afternoon
awāwa—Valleys
‘ekahi—One, first
‘ele‘ele—The darkened clouds piled high
etak—Micronesian system of navigational triangulation using island alignment with stars and swell patterns to determine dead-reckoning position
hā—Breath
ha‘a—Hawai‘i dance with bent knees
ha‘ano‘u—Hawai‘i vocal quality of loud, forceful articulations contrasted with soft, gentle sounds; see *aheahe*
haehae—To tear or rip
hāhā—To feel, grope
ha‘i kupuna—Hawai‘i genealogical chant
ha‘i mo‘olelo—Hawai‘i storytelling
ha‘i‘ōlelo—Oration, speech
haka—Empty space; a recipient as a medium or oracle; to stare; Hawaiian Star Compass house that refers to the emptiness of space near the celestial poles because of the lack of stars in the region

haku—to compose, invent, arrange, braid, plait, or put words in order; the head of household loyal to the *'ohana*
haku mele—Hawai'i composer or composition of poetic expression
haku mo'olelo—Hawai'i composer or composition of historical records
hakuone—Land area cultivated by the *maka 'āinana* for the *haku*
hakupa'a—A new *kalo* patch
hala—The pandanus tree
hālala—Overly large
Hālala—The name of Kalākaua's *ma'i*
hālau—Hawai'i place of learning
hālau hula—Hawai'i place of learning Hawai'i dance
hālāwai—Horizon
hale—Hawai'i house
Hāloa—Second son of Wākea and Ho'ohōkūkalani
Hāloanakalaukapalili—Premature son of Wākea and Ho'ohōkūkalani
hānau—To give birth
hanu—To breathe, respire, transpire
haumana—Student
Haumea—The divine entity associated with life-forming processes on land and in the sea
he oli kīpaepae—A request to enter
hea—Misty, clouded, obscured
heiau—Stone temples or places of worship; space/time where the movement of energy can be captured
helu—To list
hema—Left, opposite of *'ākau*
Hi'iaka—General name for many of Pele's younger sisters but usually referring to Hi'iakaikapoliopele
Hi'iakakuilei—Hi'iaka entity associated with the alignment of cones, mountains, and islands, much like footsteps in the sand; represents the continuum of volcanic activity
Hi'iakaikapoliopele—Youngest sibling of the Pele clan and an entity associated with generating life on the barren lava fields
Hi'iakakalukalu—Hi'iaka entity associated with the thinning of lava as it spreads across the surface of the earth
hiki—To arrive
hikina—Form of the term *hiki*
hina—To lean or fall from an upright position
Hina—Entity associated with female fertility and the innate nature in every entity to flourish, thrive, and multiply

Hinapuku'ai—Hina entity from whom vegetable foods emerge
Hinapukui'a—Hina entity from whom fish emerge
Hinaulu'ōhi'a—Hina entity associated with the *'ōhi'a* forest growth
hiwi—Skinny, bony, thin, angular; sharp ridge of a mountain
hō'ae'ae—Hawai'i chant genre with lengthened vowels; patterned and contoured use of sustained pitches
ho'ānu'unu'u—Hawai'i vocal quality of quick trilling pulse-like sounds
ho'ā'o—To try or taste
hō'emi—Waning moons
hohoa—Round *kapa* beater
hohola—Spread out, unfold, unfurl
hōkū—Star
hōkū kia'i—Guiding or guardian stars used by navigators
hōkū no ka malama—Stars for every month of the year
hōkū no ke akua—Stars relating to the divine entity
hōkū'ae'a—Wandering stars, also known as planets
hōkū'āina—Land stars
hōkūali'i—Royal stars
hōkūhō'ike—Prophetic stars
hōkūkahuna—Stars for the Hawai'i priests
hōkūkilo—Stars observed by Hawai'i astronomers
Hōkūle'a—*Wa'a kaulua* built in the 1970s by the Polynesian Voyaging Society to test the theory that purposefully planned long distance non-instrument ocean navigation was possible
hōkūlewa—Moving stars
hōkūlewa 'ano 'ole—Insignificant moving stars
hōkūmaka'āinana—Plebian stars
hōkūpa'a—Fixed stars
Hōkūpa'a—The fixed star also known as Polaris
honi—To smell, sniff
honua—The physical entity known as the earth
Honuaiākea—Canoe built by Pele for her journey from Polapola to Hawai'i, whose name describes the expanse of earth
ho'o—Causative prefix, causes the verb that follows
Ho'ohōkūkalani—Daughter of Wākea and Papahānaumoku, who creates the stars and initiates spiritual illumination
ho'oilo—Wet season
ho'okele—Hawai'i navigation
ho'oku'i—Zenith

ho'olohe—To hear; describes how the mind/body imbues what is heard with meaning
ho'olua—Northwest quadrant of the Hawaiian Star Compass named for the strong north wind associated with storm systems passing the islands
ho'onui—Waxing moons
ho'opā—To touch
ho'opa'a—To make fast, persist, memorize; also refers to the hula drummer and chanter (those who memorize the accompaniment of the *hula*)
ho'oponopono—Hawai'i practice of forgiveness and reconciliation
ho'ouwēuwē—Hawai'i chant genre of funerary wailing
hope—Space/time in back of body; backward, in back of, after; the stern of a canoe
hula—Hawai'i dance
hula ali'i—Hawai'i dance type honoring a person of high political leadership status
hula 'auana—Hawai'i dance form reflecting more modern rhythms, movements, and adornments
hula 'ili'ili—Hawai'i dance type featuring the use of smooth, flat, hand-sized stones or pebbles
hula 'īlio—Hawai'i dance type featuring the movements of a dog
hula kahiko—Hawai'i dance form reflecting more traditional rhythms, movements, and adornments
hula mu'umu'u—Hawai'i dance type featuring the movements of a maimed person, usually performed while kneeling
hula pahu—Hawai'i dance type featuring the use of a drum
hula pa'i umauma—Hawai'i dance type featuring chest slapping
hula Pele—Hawai'i dance type that honors Pele
hula pua'a—Hawai'i dance type featuring the movements of a pig
hula wahi pana—Hawai'i dance type honoring a legendary, historic, celebrated place
huli—To search for, to research, to study
i ka wā kahiko Hawai'i—In old time Hawai'i , usually referring to pre-European contact
i ka wā ma hope—The space/time in back of your body
i ka wā ma mua—The space/time in front of your body
i ka wai hū—Gushing up from springs
i ka wai kau—Aqueducts
i lalo i ka honua—Deep in the ground
iho—Downward, on bottom of, below, under
ika—The sides of a *mala* where grass is thrown
ikaika—strong

ʻike—To know or understand, recognize, perceive; describes how the mind/body imbues what is seen with meaning

ʻili—Land division within an *ahupuaʻa*; skin

ʻili ʻāina—Type of *ʻili* assigned to warrior chiefs who shared directly with the *aliʻi ʻai ahupuaʻa*

ʻili kūpono—Type of *ʻili* assigned to *kaukau aliʻi*, the yields from which were shared directly with the *aliʻi nui ʻai moku*. Leadership of this land area did not change regardless of changes in leadership at the *ahupuaʻa* level.

ʻili lele—Type of *ʻili* consisting of several distinct land sections, whose edges often did not touch each other yet formed one unit

ʻiliʻili—Small smooth flat stones

ʻilima—Dry shrublands

ilo—To germinate

inoa—Name

ipu hōkeo—Long gourd bowl, generally used to hold food, clothing, fishing gear

iwi—The side of a dryland *kalo* field

ka ʻōlelo mua—Introduction

ka lā hiki—East, poetic term

ka lā kau—West, poetic term

kaʻao—Traditional story, tale of ancient times, legend or myth

kahakai—Coastal region

kahaone—Sandy beach

kahawai—Freshwater stream

kāhela—Gourd drumbeat where the gourd is thumped down on a pad then raised and slapped twice with the fingers

kahi—Place

kahiki—In context of celestial physiology, the arrival of celestial bodies

kahikikapapalani—Part of the celestial sphere where celestial bodies appear to slow across the upper echelons of the sky

kahikikapapanuʻu—Part of the celestial sphere where celestial bodies appear to quickly ascend

kahikikapuihōlaniikekuina—Part of the celestial sphere where celestial bodies are nearest the apex of the sky

kahikikū—Part of the celestial sphere where celestial bodies appear to stand upright without the support of the earth and no longer appear to be traveling on a slanted path

kahikimoe—Part of the celestial sphere where celestial bodies appear to be lying on the surface of the earth, slowly rising from sleep

kahikole—Early morning

kahikū—Midmorning
kāholo—The "vamp" hula step, more common in modern than in ancient dance
kahuna—Hawaiʻi master practitioner, grounded in a particular genealogy and selected to complete several levels of advanced training. These advanced levels of knowledge are guarded and specific to each practitioner's genealogy.
kāhuna—Plural form of *kahuna*
kahuna hāhā—Type of medical professional capable of making diagnoses such as sickness or pain simply by feeling or groping the body
kai—Sea, sea water, area near the seaside
kai ʻelemihi—Sea area associated with a type of crab
kai heʻenalu—Sea area associated with surfing activity
kai kāhekaheka—Sea area associated with small salt-collecting pools
kai kāʻili—Sea area associated with hook and line fishing without a pole
kai kiʻokiʻo—Sea area associated with water-collecting rocky basins
kai paeaea—Sea area associated with pole fishing activity
kaiao—Dawn
Kaiwikuamoʻo—Star line family known as lizard backbone
kakahi—Solitary, unique, outstanding
kakahiaka—Morning
kākāʻōlelo—Hawaiʻi oratory
kālaiʻāina—To manage or direct the human affairs of the land
kālaimoku—Chief counselor to the *aliʻi*
kalana—Land section similar to a *moku* on the island of Maui
kalanipaʻa—The fixed firmament, or celestial sphere
kalo—Taro
Kalupeakawelo—Star line family known as the kite of Kawelo
kāmākua—Hawaiʻi genealogy and origin chants
Kamohoaliʻi—Pele clan entity associated with locating potential sites for volcanic activity
to occur
kanaka—Human being; man, person, or individual
kānaka—Plural form of *kanaka*
Kanaka Hawaiʻi—Hawaiʻi persons
Kanaka Maoli—Full-blooded Hawaiʻi person
Kānaka ʻŌiwi Hawaiʻi—Native persons of Hawaiʻi genealogical lineage
Kanaloa—Divine entity most commonly associated with the elements and processes of the ocean, but also associated with those located on land in freshwater aquifers, rift zones, pumice, and estuaries

kāne—Male, man, husband
Kāne—Entity associated with life-giving processes such as freshwater, sunlight, and air
Kānehekili—Pele clan Kāne entity associated with the thunder that rumbles from within the earth
Kānehoalani—Kāne entity associated with the life-giving processes of the sun
Kāneikōkala—Pele clan Kāne entity associated with volcanic dykes
Kānekālaihonua—Canoe built by Pele for her journey from Polapola to Hawai'i, whose name describes the Kāne entity who carves earth
Kānemiloha'i—Pele clan Kāne entity associated with lava tubes that originate from deep in the earth
kanikau—Hawai'i lament, a mourning song
Kanilehua—Mistlike rain that quenches the thirst of *lehua* blossoms associated with Hilo
Ka'ōahi—Female leader of Ni'ihau who Pele intended to visit on her journey from Polapola to Hawai'i
kāohi—Hawai'i vocal quality of cutting off prolonged vowel sounds
kaona—Multiple meanings
kapa—Bark cloth
Kapōkohelele—Pele clan entity associated with spreading of lava on the surface of the earth
Kapō'ulakīna'u—Pele clan entity associated with the direction lava will take on the surface of the earth
kapu—Prohibition, sacred or restricted
kau—To place, set down, or to rest; a place or period of time, a season, specifically summer season; a sacred prayer or chant of affectionate greeting to persons, hills, and landmarks; a chant of sacrifice to a deity; to chant
kau ka lā i ka lolo —Noon
kau lana ka lā—The floating sun
Kauilanuimākēhāikalani—Pele clan entity associated with lightning that originates from the earth, not the sky
kaukau ali'i—Advisers
kauwela—Hot season
kāwele—Hawai'i chant genre with distinct pronunciation somewhat like *kepakepa* but slower
ke ala ula a Kāne—The flaming path of Kāne, the eastern sky
ke alanui i ka piko o Wākea—The great path of the navel of Wākea; the celestial path of the sun at the equinox
ke alanui ma'awe a Kanaloa—The wispy great path of Kanaloa, the western sky

ke alanui polohiwa a Kanaloa—The large dark glistening path of Kanaloa; the celestial path of the sun at the winter solstice
ke alanui polohiwa a Kāne—The large dark glistening path of Kāne; the celestial path of the sun at the summer solstice
Ke ali'i o kona i ka lewa—Star also known as Canopus, the second-brightest star in the sky.
Kealaikahiki Channel—Ocean passage between Lāna'i and Kaho'olawe that leads to Kahiki
keiki 'alu'alu—Premature child
Kekāomakali'i—Star line family known as the bailer of Makali'i
kepakepa—Hawai'i chant genre of rapid conversational patter, where every syllable is clearly pronounced without prolonged vowels or sustained pitches, thereby not requiring too much breath
kīhāpai—A garden, field, or small farm
kinolau—Many bodies
kiopa'a—The north star
kīpaepae—Stone pavement or steps for entering a house
Kīpu'upu'u—Thick, heavy, drenching cascade of the pelting rain associated with Waimea
kīpuka—A variation of change in form, such as an oasis within a lava bed
ko'ele—Land area cultivated by the *maka 'ainana* for the *konohiki*
ko'ihonua—Hawai'i genealogical chant
kole—Red, swollen, inflamed
komo—To enter
Komo'awa—Kahuna to Wākea
komohana—Form of the term *komo*
kona—South; older term based on wind pattern
Kona—Southwest quadrant of the Hawaiian Star Compass, named for the leeward side of the island chain where it is protected from or in the lee of the constant blowing of the "trade" winds
konohiki—Land steward responsible for maintaining productivity of an area by managing people's consumption of and use relationships with water, land, agriculture, and aquaculture within a land area known as an *ahupua'a*; this person was loyal to the *ali'i 'ai ahupua'a* who maintained the land and nearshore day-to-day activities
ko'olau—North; older term based on wind pattern
Ko'olau—Northeast quadrant of the Hawaiian Star Compass, named for the windward side of the island chain where the "trade" winds originate
kū—To rise, stand erect

Kū—Divine entity commonly associated with war, governance, taking action, and making critical decisions, as well as male fertility, but also associated with the intangible essence in every entity to grow, adapt, and evolve; the dry, hot season, *kauwela*

kua—Back, spine; on a islandscape, the mountain region

Kua—Shark, considered an ancestor for many families from Kaʻū and namesake for Kealakekua

kua laʻau—Wooden anvil

kua pōhaku—Stone anvil

kuahea—Mountain region where mist forms and trees are stunted by altitude

kuahiwi—Mountain region extending from the *kualono* to where the land flattens

kuahu—Altar

kuakua—A cultivated thin strip of land on the embankment between taro patches

kualapa—Mountain ridge

kualau—Wind-driven rain

kualipi—Sharp mountain ridge with a thin tapering edge like that of an axe

kualono—Mountain region where the sounds of silence are deafeningly loud

kuanihi—Steep mountain ridge with difficult or precarious passage

kuaola—Verdant mountain region where all things grow and thrive, as *ola* is life

kuāuna—Valley-like stream bank

Kuehulepo—Dust-stirring winds associated with Kaʻū

Kūhaʻimoana—Pele clan entity associated with lava tubes that span the ocean floor

Kūholoholopali—Kū entity associated with moving a prospective canoe hull down steep elevations

kuʻialua—Hawaiʻi martial arts

kuina—A meeting place

kuʻina—A connective joining

Kūkaʻōhiʻalaka—Kū entity associated with the *ʻōhiʻa* tree who is also connected with Laka

kūkū—Gourd drumbeat where the gourd is thumped down on a pad three times then raised and slapped twice with the fingers

kukui—Candlenut; a large nut-bearing tree with oily kernels that have many uses but were used mainly for light; the tree became associated with enlightenment. The cooked nuts were used as a relish, the cooked shell and gum from the bark and the roots as a dye. It is also the *kinolau* of Kamapuaʻa (a pig demigod associated with Lono).

kūkulu—A point or end of the earth

kūkulu ʻākau—North, based on sun movement

kūkulu hema—South, based on sun movement

kūkulu hikina—East, based on sun movement
kūkulu komohana—West, based on sun movement
kula—Dry vegetation lands
Kulāiwi—Homeland; Oliveira's sense of place/time lineage, link between ancestral knowledge and place
kuleana—Privileged responsibility, privileged burden
kūmākena—To lament, bewail, mourn loudly for the dead
Kūmokuhāli'i—Kū entity associated with the spreading land
kumu—Hawai'i teacher, instructor, mentor
kumu hula—Hawai'i dance teacher
Kūpā'aike'e—Kū entity associated with the stone-devouring bent-shaped adze
Kūpulupulu—Kū entity associated with tinder, chip, mulch, kindling maker
kupuna—Hawai'i elder
kūpuna—Plural of *kupuna*
Kū'ulakai —Kū entity associated with the abundant sea and revered by fishermen
Kū'ulauka—Kū entity associated with the abundant inland, also known as Kū the cultivator
lā—The physical entity known as the sun
Lā—Hawaiian Star Compass house nearest the equator that represents the region where the sun spends most of its time throughout the year
lā'au lapa'au—Hawai'i practice of addressing pains and illnesses with plants and prayer
Laka—Divine entity associated with hula
lalani—The Milky Way
lalo—Down
lama—An endemic ebony hardwood tree, a torch or light
lani—The physical entity known as the sky
lapa—A steep side of a ravine
Leho—Entity potentially associated with the processes that create sulfur-crusted earth
lehu—Ash
lehua—Flower of the *'ōhi'a* tree and also the tree itself, generally associated with the Island of Hawai'i
lehulehu—Multitude
lei—Hawai'i garland of flowers
lepo—Soil
lewa—The lower regions of the skyscape where atmospheric bodies float, dangle, oscillate, or are otherwise suspended
lewaho'omākua—Atmospheric layer where many bodies continue to multiply, thrive, and flourish

lewalani—Highest atmospheric layer where a majority of atmospheric bodies, both entities and processes, reside, such as the various winds, rains, and cloud formations
lewalanilewa—Atmospheric layer where low-hanging clouds move across the sky and occasionally kiss the surface of the highest mountains
lewanuʻu —Atmospheric layer where birds fly
lihi kai—Sea water edge
limu—Seaweed
limu koko—A variety of seaweed
lipi—A thin tapering edge, like that of an axe
lohe—To listen; describes the ear's physical capability
loʻi—A terraced irrigated pond
loʻi kalo—Taro ponds
loko—Inside of, within
lomilomi—Hawaiʻi massage
lono—To hear
Lono—Divine entity associated with agricultural fertility, peace, and recreation, as well as the wet season, *hoʻoilo*
Lonomākua—Lono entity who presides over *makahiki*; also associated with being the fire keeper, a revered relative of Pele
lua—Crater, pit
luakini heiau—A place of worship of the highest order
luna—Up
māhū—Transgendered class of people
māhuahua—Hawaiʻi *meiwi* involving reduplication or repetition of syllables
māhūkāne—A biological female who is man-like
māhūwahine—A biological male who is woman-like
mai—Toward the body
maʻi—Genitals
maile—A fragrant native twining shrub with shiny leaves
Māʻilikūkahi—An Oʻahu *mōʻī*, with the establishment of *palena ʻāina* on Oʻahu
maka—Eye, beginning, point of a spear
makaʻāinana—General populace or citizens
makahiki—A year; a celebration marking the start of the Hawaiʻi year
makai—Oceanward, toward the ocean
Makaliʻi—Star family also known as the Seven Sisters or the Pleiades
makana—Gift
makawalu—An infinite movement, evolution, and transformation of the eight eyes; a metaphor or symbol of chiefly *mana*, which is considered the most supreme form of *mana* for a human

mākua—Full grown, established; to be large, grow, strengthen, sustain
mala—Cultivated dryland garden
malama—Month; the physical entity known as the moon
malama pili—Related or coinciding months
malanai—Southeast quadrant of the Hawaiian Star Compass, named for the gentle breeze associated with the southeast part of O'ahu and Kaua'i
malua—A small hill dug for planting potatoes
mana—Power, strength, might, supernatural power, spirit, energy of character, glory, majesty, intelligence; a line projecting from another line like a branch of a tree or of a stream; a variant or version as of a story; a limb of the human body and the stage of growth in which a fetus forms limbs; the stage of growth of a fish in which colors appear; the name of a specific house on a *luakini heiau,* a place of worship of the highest order; a variety of *kalo,* taro, used as medicine that propagates by branching from the top of the corm; a native fern with several large subdivided fronds; and a particular type of fishhook used to catch eels
Mānaiakalani—Star family named after Maui's fishhook, also known as Scorpio
manawa—Both a period of time and a space or interval between two objects or events; the fontanel
manō—Shark
manu—Bird
Manu—Hawaiian Star Compass house that metaphorically represents the canoe as it travels on the ocean above the marine life, much like a bird flies in the atmosphere above the terrestrial life
Maui—Divine entity that represents intellectual inquiry, ingenuity, and action
mauka—Toward the mountain
mauna—Mountain
mea haku—One who composes
meiwi—Literary devices such as mnemonics, repetition, rhythm, and rhyming schemes
mele—Hawai'i poetic oratory
mele aloha—Hawai'i poetic oratory of love
mele aloha 'āina—Hawai'i poetic oratory expressing deep affection for the land
mele ho'āla—Hawai'i poetic oratory composed to wake a sleeping child
mele ho'i—Hawai'i poetic oratory for processional entry
mele ho'ohanohano—Hawai'i poetic oratory glorifying a chief
mele ho'oipoipo—Hawai'i poetic oratory of lovemaking and wooing
mele hula—Hawai'i poetic oratory meant to accompany hula
mele inoa—Hawai'i poetic oratory honoring the name of a person or place
mele kāhea—Hawai'i poetic oratory for processional entry

mele ka'i—Hawai'i poetic oratory for processional entry
mele ko'ihonua—Hawai'i poetic oratory expressing cosmogonic genealogy
mele komo—Hawai'i poetic oratory for processional entry
mele kū'ē—Hawai'i poetic oratory of resistance
mele kūmākena—Hawai'i poetic oratory of lament
mele ma'i—Hawai'i poetic oratory honoring the life-giving potential of the genitalia of a specific person, usually an *ali'i*
mele nema—Hawai'i poetic oratory criticizing or challenging others
mele oli—Hawai'i poetic oratory not meant for dancing, see *oli*
mele pai ali'i—Hawai'i poetic oratory honoring a chief by name-dropping
mele pana—Hawai'i poetic oratory of a celebrated or noted place
moe—To sleep
Moemoea'ali'i—Entity associated with dormancy and the latency or potentiality of life
mō'ī—Supreme leader; Beamer suggests a possible translation of "a succession of the supreme"
moku—Island, district, land division
mokupuni—Island
mo'o—Any kind of reptile such as lizard, dragon, serpent or water spirit; a succession or series, especially a genealogical line; a story, tradition, legend, though this usage is less common than *mo'olelo*; a narrow strip of land, smaller than an *'ili*—also *mo'o 'āina*; a small fragment, as of cloth; a narrow path, track such as the raised surface extending lengthwise between irrigation ponds; a mountain ridge; young, as of pigs, dogs; grandchild; brindled, as a dog; streaked, tawny, as cattle; color of a tabby cat; *ho'omo'o*—to follow a course, continue a procedure; on a canoe it is the gunwale strakes or side planks fitted to the middle section on each side of a canoe hull
Mo'o—Oliveira's sense of being the right person in the right place and the right time to whom ancestral knowledge flows
mo'okū'auhau—Genealogy
mo'olelo—Hawai'i historical accounts
mua—Space/time in front of body; forward, in front of, before; on a canoe it is the bow
muliwai—Freshwater stream mouth
nā akua—Multitude of divine entities or processes
nā alanui o nā hōkū ho'okele—The great paths of the navigation stars used to guide the canoe across the ocean
nā ali'i—Hawai'i political leadership class
nā hōkū i ka lewa a me ka lipo—The stars of the sky and the darkness

nā hōkū pa'a o ka 'āina—The fixed stars of the lands that are considered celestial markers for specific island groups; for example, Hōkūlea is the star over the islands of Hawai'i

nā kāhuna—Hawai'i class of master practitioners

nā leo—The voices

Nā Leo—Hawaiian Star Compass house that refers to the voices of the stars speaking to the navigator and acknowledges the intimate communication Kanaka Hawai'i have with nature

na'au—Small intestines and, metaphorically, the seat of thought, intellect, affections, and moral nature

Na'au—Oliveira's sense of ancestral lineage, intuition

nahele—Wilderness

Nālani—Hawaiian Star Compass house associated with the second-brightest star in the sky, *Ke ali'i o kona i ka lewa*, also known as Canopus

Nāmakaokaha'i—Pele clan entity associated with geological faults or weak points in the earth

nānā—To look at something, to observe it, pay attention to it, or take care of it; describes the eye's physical capability

nānā i ka wā ma mua—Look to the space/time in front of you

nei—Space/time chameleon with many meanings depending on what it accompanies; used as a present tense verb marker; used as a term of endearment with nouns or pronouns; used to indicate past time with directional terms

Newe—The star family also known as the Southern Cross

nihi—A steep edge or border with difficult or precarious passage

niho pōhaku—Tooth-stone used to build interlocking rock foundations

Noio—Hawaiian Star Compass house named after the Hawai'i noddy tern, a small black bird that lives on fish and helps a navigator find islands because it flies out to sea in the morning and returns to land at night to rest

nu'u—High place, crest, summit; to rise up, be full or high

nui—Great or large

'ō aku—On the far side of/beyond the body

'o ia ko'u hope iho—S/He is before or below me

'o ia ko'u mua a'e—S/He is before or above me

'ohana—Family, relative, kin group, related, a group of kindred individuals, lineage, race, tribe, those who dwell together

'ohe kāpala—Bamboo stamps used for making *kapa* designs

'ōhi'a—A type of large tree in the Hawai'i forest

'ōiwi—Native

ʻokana—Land division smaller than a *moku* but larger than an *ahupuaʻa*
ola—Life
ʻōlelo nane—Hawaiʻi riddle, parable, allegory
ʻōlelo noʻeau—Hawaiʻi proverb
oli—Chant not accompanied by a hula
oli kāhea—Recital of the first lines of a stanza by the dancer as a cue to the chanter; chant performed to enter a *hālau hula*
olioli—Hawaiʻi chant genre of a sustained monotone pitch with embellishments using upper and lower neighbor tones
ʻoʻopu—A goby fish that lives in brackish water and is considered a delicacy for its sweet succulent meat
ʻōpaka—Mountain ravine
pā—To touch; gourd drumbeat where the gourd is thumped down on a pad then raised and slapped once with the fingers
pae ʻōpua i ke kai—Cumulus or billowy, puffy clouds banked up near the horizon
paha—Improvised or conversational chants
pahu—Drum
pālaha—Spread out, broaden, flatten
palapalai—A native fern
palena—Border, boundary, or dividing line between two places, but might also be explained as a protected place
palena ʻāina—Land boundary
pali—Mountain precipice or cliff
palu—Micronesian term for a master navigator
panopano—Thick black clouds
paoa—A rare variation of *pawa*
papa—Foundation, stratum, flat surface, layer
papa kahuna pule—Master priests class
Papahānaumoku—Divine entity considered the foundation that births islands; female progenitor of Hawaiʻi society papahulihonua—Knowledge stratum studying the earth and ocean including its development, transformation, and evolution by natural causes
papahulilani—Knowledge stratum studying the sky, the space above our heads to where the stars sit in the atmospheric expanse
papakū—An established foundation of both tangible and intangible entities bound together and standing as one
Papakū Makawalu—Foundation of constant growth, a Hawaiʻi discourse established by Pualani Kanakaʻole Kahahele
papanuihānaumoku—Knowledge stratum studying the birthing processes of all flora and fauna including humans

pāʻū—Skirt
paukū—A small lot of land
pawa—Pre-dawn, when the light from the sun first breaches the horizon and lightens the darkness
pele—Magma
Pele—Both a person and a divine entity associated with volcanic landforms and processes
Pelehonuamea—Another name for Pele
piha poepoe—Round or full moons
piko—End or extremity of an entity, such as the end of a rope, navel, or belly button
pili—Near in proximity
pilipuka—After midnight, near morning
pīnaʻi—Hawaiʻi *meiwi* involving a repeated term
pō—Darkness, night, obscured
pō mahina—Moon phases
Pōʻaiʻoluʻākau—Tropic of Cancer
Pōʻaiʻoluhema—Tropic of Capricorn
Pōʻaiwaenahonua—Equator
poho—A depression, hollow, cupule
Pōkāneloa—Solar clock rock on the island of Kahoʻolawe
pōlehulehu—Dusk
Poliʻahu—Divine female entity associated with snow
pono—Proper, righteous, balanced with regard to give and take
poʻopuaʻa—Head student of a *halau hula*, so named because they tend to the *kuahu*
pōpolohua—The purplish-blue reddish-brown clouds
pua—Flower
puaʻa hiwa—Black pig
puʻeone—Sandbar
puka—A hole; or to pass through or appear
pule—Prayer
Pwo—Micronesian term for a sacred master of both the technical and spiritual aspects of Micronesian non-instrument ocean navigation
uakoko—Red rainfall
ʻuala—Sweet potato
uē helu—Hawaiʻi chant of lament
ʻukulele—Hawaiʻi instrument of Portuguese origin
wā—Spatial/temporal interval
wā ʻekahi—First space/time era or interval

wā ʻelua—Second space/time era or interval
waʻa—Canoe
waʻa kaulua—Long-distance double-hulled voyaging canoe
wae—To choose or select
waena—Between, in the middle of, among, in the center of
wahi pana—Celebrated place
wahine—Female, woman, wife
waho—Outside of
wai—Fresh water
wai kau a Kāne me Kanaloa—The aqueducts of Kāne and Kanaloa
Wākea—Divine entity considered the broad expanse of the sky; male progenitor of Hawaiʻi society
wana—To appear
wanaʻao—Dawn
wao—Wet forest region
wao akua—Wet forest region for the *akua*
waoʻamaʻu—Wet forest region where ferns grow
waoʻeiwa—Wet forest region covered with vegetation and small forest trees
waokanaka—Wet forest region for Kanaka Hawaiʻi
waolipo—Wet forest region with dense vegetation and tall forest trees
waomaʻukele—Wet forest rain belt where the ground is always wet and slippery
waonahele—Inland wet forest region covered in vegetation
wauke—Paper bark
wela—Hot

Notes

1 In Hawai'i, before any significant undertaking it is necessary to ask permission. According to *kumu hula*, Hawai'i dance teacher, and Hawai'i scholar Pualani Kanaka'ole Kanahele, in *Holo Mai Pele* (the book that accompanies the performance created by her and her sister, Nalani Kanaka'ole, and performed by Hālau o Kekuhi), "Kūnihi ka Mauna" is a *mele oli*, Hawai'i poetic oratory not meant for dancing, often used to ask permission to enter into a *hālau*, Hawai'i place of learning. Ku'ualoha Ho'omanawanui, Hawai'i literary scholar, uncovered five versions of this chant published in nineteenth-century newspapers and at least seventeen unpublished versions in various archival collections. She notes that "although the individual contexts of each of these versions vary, they all express the traditional Hawaiian attitude of not 'barging in' where one does not have *kuleana*, privileged responsibility, and of showing respect for the people, place, and culture where one is an outsider" (Ho'omanawanui 2002). I have frequently used this *mele* at the beginning of conference presentations or workshops to metaphorically ask permission to enter a symbolic place of knowledge exchange. I do it specifically to honor my ancestors as well as the ancestors dwelling in the place where the conferences/workshops are located and to set the tone of the presentation. I have included this *mele oli* here for similar reasons.

2 The use of the term "Western" in this text is best explained as a compilation of Euro-American perspectives.

3 The use of the term "Indigenous" in this text is best explained as a compilation of Native perspectives.

4 See M. Kovach, *Indigenous Methodologies: Characteristics, Conversations and Contexts* (Toronto: University of Toronto Press, 2009); Research Regulation in American Indian/Alaska Native Communities: Policy and Practice Considerations. NCAI Policy Research Center Webpage.

5 See J. Abdullah and E. Stringer, *Indigenous Knowledge, Indigenous Learning, Indigenous Research* (New York: Falmer Press, 1999); R. Bishop and T. Glynn,

Culture Counts: Changing Power Relations in Education (London: Zed Books, 2003); C. Crazy Bull, "Advice for the Non-Native Researcher," *Tribal College: Journal of American Indian Higher Education* 9 (1): 25–27; B. Harrison, *Collaborative Programs in Indigenous Communities: From Fieldwork to Practice* (Walnut Creek, CA: AltaMira Press, 2001); A. Heiss, "Writing about Indigenous Australia: Some Issues to Consider and Protocols to Follow: A Discussion Paper," *Southerly* 2002; M. Ivanitz, "Culture, Ethics and Participatory Methodology in Cross-Cultural Research," *Australian Aboriginal Studies* 2 (1999): 46–58; J. A. Kievit, "A Discussion of Scholarly Responsibilities to Indigenous Communities," *American Indian Quarterly* 27 (2003): 1–2; D. A. Mihesuah, *Natives and Academics: Researching and Writing about American Indians* (Lincoln: University of Nebraska Press, 1998, xi); R. A. and D. Deur, "Reciprocal Appropriation: Toward an Ethics of Cross-Cultural Research," in *Geography and Ethics,* eds. J. D. Proctor and D. M. Smith, 237–250 (London: Routledge, 1999); E. Steinhauer, "Thoughts on an Indigenous Research Methodology," *Canadian Journal of Native Education* 26 (2): 69–81.

6 Kamehameha Publishing published *Clouds of Memories* in the summer of 2006.

7 Hawaiʻi language leaders did create a term, *pālakiō*, where *pā* is a prefix for "having the quality of" and *lakiō* is "ratio," to aid in mathematical instruction.

8 Kaulana Mahina is available online at the Kamehameha Publishing website, complete with information for planting and fishing and the names of the months (http://www.kpublishing.org/_assets/publishing/multimedia/apps/mooncalendar/index.html).

9 "Hula step danced with bended knees; the chanting for this dance is usually bombastic and emphatic; to dance thus" (Pukui and Elbert, 2003).

10 I've always thought it strange that Wākea breaks his own *kapu*. Though the story of Adam and Eve similarly speaks of heightened intellect occurring from breaking rules, in this story it is the male who is the one who lures a seemingly innocent and naive young woman to break the rules.

11 The formal title is *Buke Kakau Paa no ka mahele aina i Hooholoia i waena o Kamehameha III a me Na Lii a me na Konohiki ana*. It is a record of the land divisions between Kamehameha III (Kauikeaouli), the *aliʻi* (chiefs), and the konohiki (person loyal to the *aliʻi ʻai ahupuaʻa* who maintained the land and nearshore day-to-day activities). The document was completed in 1848.

12 Manu Meyer introduced this term to me.

13 Thomas Samuel Kuhn (1922–1996) is one of the most influential philosophers of science of the twentieth century. His book *The Structure of Scientific Revolutions*, published in 1962, is an analysis of the history of science. Up to that point, philosophers viewed scientific progress as cumulative. Kuhn argued that

there were periods of growth that were interrupted by periods of revolutionary science where the "anomalies" normally set aside in scientific inquiry lead to a whole new direction of understanding. In other words, science evolves as a result of changing intellectual circumstances and possibilities.

14 Although the settlement of the Islands of Hawai'i had been widely recognized as occurring between 300–750 CE, high-precision radiocarbon dating indicates human settlement of the Eastern Polynesian Islands was more recent, remarkably rapid, and extensively dispersed. As a result, the new estimates of human settlement of the Islands of Hawai'i are between 1190–1290 CE (Wilmshurst et al. 2011).

15 See Finney 1998 for more information about Caroline Island navigation and cartographic mentoring processes.

16 The details of these star line families are also accessible online at the Polynesian Voyaging Society website, www.hokulea.com. I am confident the information and the website will be available for many years.

17 For a more complete and comprehensive story of this Polynesian voyaging renaissance, read *Hawaiki Rising* by Sam Low.

18 Cognitive cartography is a dynamic, flexible, and internal process "covering those cognitive or mental abilities that enable us to collect, organize, store, recall, and manipulate information about the spatial environment" (R. M. Downs and D. Stea, *Maps in Minds: Reflections on Cognitive Mapping*, New York: Harper and Row, 1977). It involves processes that require interaction with external stimuli so we can adapt, augment, or otherwise modify our stored spatial /temporal knowledge and stay attuned with the dynamic, fluid, and rhythmic nature of the world around us.

19 Cognitive processes are generally understood as including "sensation and perception, thinking, imagery, memory, learning, language, reasoning, and problem-solving" (Smelser and Bates, *International Encyclopedia of the Social and Behavioral Sciences*, Oxford: Pergamon Press, 2001), 14771. Spatial cognition involves the acquisition, organization, utilization and revision of knowledge about real and/or abstract spatial/temporal environments and includes both "knowledge *and beliefs* about spatial properties of objects and events in the world." "Spatial properties of the world include location, size, distance, direction, shape, pattern, movement, and inter-object relations" (D. R. Montello, 1997, Unit 006 - Human cognition of the spatial world. Retrieved from http://www.ncgia.ucsb.edu/education/curricula/giscc/units/u006/u006.html).

20 *Hōkūle'a* does not have a keel or a daggerboard and cannot sail into the wind. It can only sail about 67 degrees port or starboard (left or right, respectively) from the direction of the wind source.

21 While at sea there is only one navigator on board a *wa'a*. Although Mau was more experienced, he was not the navigator for this voyage.

22 During the summer solstice the canoe would be two points forward from the port beam at sunrise. During winter solstice, the canoe would be six points forward from the port beam at sunrise.

23 In summer the canoe would be two points back from the starboard beam; in winter it would be six points back from the starboard beam; and at the equinox it would be four points back from the starboard beam.

24 My metaphoric understanding of "Hilo Hanakahi" is that it is about a lover seeking to quench her thirst and, aroused by the heady fragrance of her lover, begins to stir up a passionate encounter causing various bodily fluids to streak across their bodies as a whispered gasp of pleasure escapes during the heavy cascade of an embrace that strengthens with repeated thrashings bringing them closer to the soaring heights of impassioned release. How close did that compare with your understanding? Keep in mind the lover in this symbolic imagery could very well be the island of Hawaiʻi itself and not a human being.

25 This is similar to the idea that while anyone can, and seemingly everyone does, make maps, especially with the plethora of online mapping apps available, not everyone is trained or mentored in the art and science of cartography. Though I must clarify that whether or not a person has formal training or mentoring in either cartography or *ka haku mele* does not necessarily make them better at their craft, as there are some people who are just naturally gifted at cartographic processes or composing Hawaiʻi verbal arts.

26 I learned about *māhuahua* from Hawaiʻi language scholar and Papahulihōnua practitioner Kuʻulei Kanahele as part of a second-year Hawaiʻi language class at Hawaiʻi Community College.

27 The terms *māʻokiʻoki*, *hāwanawana*, *ʻĀpaʻapaʻa*, and *Kīpuʻupuʻu* are *māhuahua*.

28 This is another term I learned from Kuʻulei Kanahele. Perreira uses the new term *ʻēkoʻa* to represent opposites.

29 This *mele* is found in the notes section written by Nathaniel B. Emerson and was originally titled "Pule Hoʻōla" (diacritical marks added by Louis), prayer for healing.

30 Note there is no capitalization for the physical entity of lava, even though it is so named because of the association with the divine entity known as Pele.

References

Abdullah, J., and E. Stringer. 1999. *Indigenous Knowledge, Indigenous Learning, Indigenous Research*. New York: Falmer Press.

Akerman, J. R. 2009. *The Imperial Map: Cartography and the Mastery of Empire*. Chicago: University of Chicago Press.

Anderson, B. 2006. *Imagined Communities: Reflections on the Origin and Spread of Nationalism*. Rev. ed. London: Verso.

Andrade, C. L. 2014. "Geography of Hawai'i?" In *I ulu i ka 'āina: Land*, edited by J. K. Osorio. Honolulu: University of Hawai'i Press.

Andrews, L. 1865. *A dictionary of the Hawaiian language to which is appended an English-Hawaiian vocabulary and a chronological table of remarkable events*. Printed by Henry M. Whitney. http://ulukau.org/elib/cgi-bin/library?c=andrew&l=en.

Andrews, L., H. H. Parker, and Hawaii Board of Commissioners of Public Archives. 1922. *A dictionary of the Hawaiian language*. Published by the board. http://ulukau.org/elib/cgi-bin/library?c=parker&l=en.

Arista, N. 2009. "Listening to Leoiki: Engaging Sources in Hawaiian History." *Biography* 32 (1): 66–73.

———. 2010. "Navigating Uncharted Oceans of Meaning: *Kaona* as Historical and Interpretive Method." *Publications of the Modern Language Association of America (PLMA)* 125 (3): 663–669.

———. 2015. "Life and Death in the Archives: Indigenous Biography, Social Genealogies of the Lahui and the 'Aina." Transcript of Conference Paper Presentation, Native American Indigenous Studies Association, Washington, DC.

Barrow, I. J. 2003. *Making History, Drawing Territory: British Mapping in India, c. 1756–1905*. New Delhi: Oxford University Press.

Basso, K. H. 1996. *Wisdom Sits in Places: Landscape and Language among the Western Apache*. Albuquerque: University of New Mexico Press.

Beamer, K. 2008. "Na wai ka mana? 'Ōiwi Agency and European Imperialism in the Hawaiian Kingdom." In Doctorate of Philosophy in Geography, PhD diss., University of Hawai'i at Mānoa, 346.

———. 2014. *No Mākou Ka Mana: Liberating the Nation*. Honolulu: Kamehameha Publishing.

Beamer, W. D. 1987. *Nā mele hula*. La'ie, HI: Institute for Polynesian Studies, Brigham Young University. Distributed by University of Hawai'i Press.

———. 2001. *Nā mele hula*. La'ie, HI: Institute for Polynesian Studies, Brigham Young University. Distributed by University of Hawai'i Press.

Beckwith, M. W. 1970. *Hawaiian Mythology*. Honolulu: University of Hawai'i Press.

Beckwith, M. W., and K. Luomala. 1981. *The Kumulipo: A Hawaiian Creation Chant*. Honolulu: University of Hawai'i Press.

Bishop, R., and T. Glynn. 2003. *Culture Counts: Changing Power Relations in Education*. London: Zed Books.

Blaut, J. M., D. Stea, C. Spencer, and M. Blades. 2003. "Mapping as a Cultural and Cognitive Universal." *Annals of the Association of American Geographers* 93 (1): 165–185.

Brown, E. 2010. "Mau Piailug, Micronesian Who Sailed by Navigating Sun and Stars, Dies at 78." *Washington Post*, Wednesday, July 21, Post Mortem.

Byrnes, G. 2001. *Boundary Markers: Land Surveying and the Colonisation of New Zealand*. Wellington, NZ: Bridget Williams Books.

Cajete, G. 2000. *Native Science: Natural Laws of Interdependence*. Santa Fe, NM: Clear Light Publishers.

Caldwell, P. 1981. "Television News Maps." In *Technical Papers of the American Congress on Surveying and Mapping*. Washington, DC: American Congress on Surveying and Mapping.

Chapin, M., and B.Threlkeld, and Center for the Support of Native Lands. 2001. *Indigenous Landscapes: A Study in Ethnocartography*. Arlington, VA: Center for the Support of Native Lands.

Chun, M. N. 2011. *No Nā Mamo: Traditional and Contemporary Hawaiian Beliefs and Practices*. Honolulu: University of Hawai'i Press.

Condis, T. 1909a. "The Hawaiian Astronomy (History of Kanalu)." *Kuokoa Home Rula*. Vol. 7, no. 23, June 4.

———. 1909b. "The Hawaiian Astronomy (Kamohoula pt. 1)." *Kuokoa Home Rula*. Vol. 7, no. 15, April 9.

———. 1909c. "The Hawaiian Astronomy (Kamohoula pt. 2)." *Kuokoa Home Rula*. Vol. 7, no. 16, April 16.

———. 1909d. "The Hawaiian Astronomy (Kamohoula pt. 3)." *Kuokoa Home Rula*. Vol. 7, no. 17, April 23.

———. 1909e. "The Hawaiian Astronomy (Kamohoula pt. 4)." *Kuokoa Home Rula*. Vol. 7, no. 18, April 30.

———. 1909f. "The Hawaiian Astronomy (Kamohoula pt. 5)." *Kuokoa Home Rula*. Vol. 7, no. 19, May 7.

———. 1909g. "The Hawaiian Astronomy (Kamohoula pt. 6)." *Kuokoa Home Rula.* Vol. 7, no. 20, May 14.

———. 1909h. "The Hawaiian Astronomy (Kamohoula pt. 7)." *Kuokoa Home Rula.* Vol. 7, no. 22, May 28.

Crazy Bull, C. 1997. "Advice for the Non-Native Researcher." *Tribal College: Journal of American Indian Higher Education* 9 (1): 25–27.

Donaghy, J. K. 2011. "The Language Is the Music: Perceptions of Authority and Authenticity in Hawaiian Language Compositions and Vocal Performance." PhD diss., University of Otago, Dunedin, NZ.

Downs, R. M., and D. Stea. 1977. *Maps in Minds: Reflections on Cognitive Mapping.* New York: Harper and Row.

Eckardt, J. 2013. "Hula Lives: Fifty Years of Renaissance and Revival through the Merrie Monarch Festival." *Mana Magazine* (March/April). http://welivemana.com/articles/hula-lives.

Edney, M. H. 1997. *Mapping an Empire: The Geographical Construction of British India, 1765–1843.* Chicago: University of Chicago Press.

Elbert, S. H., and N. Mahoe. 1970. *Nā mele o Hawaiʻi nei: 101 Hawaiian Songs.* Honolulu: University of Hawaiʻi Press.

Emerson, N. B. 1997. Rev. ed. *Pele and Hiiaka: A Myth from Hawaii.* Honolulu: ʻAi Pōhaku Press.

———. 1998. *Unwritten Literature of Hawaii: The Sacred Songs of the Hula.* Honolulu: Mutual Publishing.

Evanari, G. K. 1995. *Polynesian Voyaging and the Wayfinding Art: A Comprehensive Curriculum Guide for Teachers and Students.* Honolulu: Self-published.

Finney, B. 1998. "Nautical Cartography and Traditional Navigation in Oceania." In *Cartography in the Traditional African, American, Arctic, Australian, and Pacific Societies,* edited by D. Woodward and G. M. Lewis, 443–492. Vol. 2, book 3. The History of Cartography series. Chicago: University of Chicago Press.

Fornander, A. 1985. *Fornander collection of Hawaiian antiquities and folk-lore: the Hawaiian account of the formation of their islands and origin of their race, with the traditions of their migrations, etc., as gathered from original sources.* Millwood, NY: Kraus Reprint (Honolulu: Bishop Museum Press, 1916–1919). (Memoirs of the Bernice Pauahi Bishop Museum of Polynesian Ethnology and Natural History, vols. 4–6.)

———. 1996. *Ancient History of the Hawaiian People to the Times of Kamehameha I (with an Introduction by Glen Grant).* Honolulu: Mutual Publishing.

Foucault, M. 1980. "Questions on Geography." In *Power/Knowledge: Selected Interviews and Other Writings, 1972–1977,* edited by C. Gordon, xii. New York: Pantheon Books.

Fox, J., K. Suryanata, and P. Hershock. 2005. *Mapping Communities: Ethics, Values, Practice.* Honolulu: East-West Center.

Fullard-Leo, B. 1997. "Chants: Mele of Antiquity." *Coffee Times*, September.

Gilmartin, P. 1981. "Influences of Map Context on Circle Perception." *Annals of the Association of American Geographers* 71 (2): 253–258.

Gon, S. O. 2010. "The Natural World of the Hula." Washington, DC: National Museum of the American Indian. Conference presentation. Archived online at http://www.ustream.tv/recorded/7308502.

Gordon, M. 2006. "New Canoe: A Tribute to Piailug." *Honolulu Advertiser*, November 14.

Handy, E. S. C., and E. G. Handy. 1972. *Native Planters in Old Hawai'i: Their Life, Lore and Environment*. Honolulu: Bishop Museum Press.

Handy, E. S. C., and M. K. Pukui. 2010. *The Polynesian Family System in Ka'u, Hawai'i*. Honolulu: Mutual Publishing.

Harrison, B. 2001. *Collaborative Programs in Indigenous Communities: From Fieldwork to Practice*. Walnut Creek, CA: AltaMira Press.

Heiss, A. 2002. "Writing about Indigenous Australia: Some Issues to Consider and Protocols to Follow: A Discussion Paper." *Southerly* 62 (2): 197–207.

Henry, T., and D. Kawaharada. 1995. *Voyaging Chiefs of Havai'i*. Honolulu: Kalamaku Press.

Ho'omanawanui, K. u. 2002. "Kūnihi Ka Mauna: The Opening Pages." *'Ōiwi* 2: iv–v.

———. 2005. "He Lei Ho'oheno no nā Kau a Kau: Language, Performance, and Form in Hawaiian Poetry." *Contemporary Pacific* 17 (1): 29–81.

———. 2013. "Ka Ola Hou 'Ana o ka 'Ōlelo Hawai'i i ka Ha'i 'Ana o ka Mo'olelo i Kēia Au Hou: The Revival of the Hawaiian Language in Contemporary Storytelling." In *Traditional Storytelling Today: An International Sourcebook*, edited by M. R. MacDonald, 160–170. 2nd ed. New York: Routledge (Gitzroy Dearborn, 1999).

Iaukea, S. L. 2012. *The Queen and I*. Berkeley: University of California Press.

Ii, J. P. 1995. *Fragments of Hawaiian History*. Edited by D. B. Barrère. Honolulu: Bishop Museum Press.

Ivanitz, M. 1999. "Culture, Ethics and Participatory Methodology in Cross-Cultural Research." *Australian Aboriginal Studies* 2: 46–58.

Johnson, R. K. K. 1993. *Kaho'olawe's Potential Astro-Archaeological Resources*. Wailuku, HI: Kaho'olawe Island Conveyance Commission.

———. 2000. *The Kumulipo Mind: A Global Heritage in the Polynesian Creation Myth*. Scribd: Self-published.

———. 2012. "The Hawaiian Understanding of the Universe." Draft doctoral dissertation, Awanuiarangi, Aotearoa.

———. n.d. "Kanikau." Retrieved from http://www.mauimuseum.org/chants.htm.

Kaeppler, A. L., E. Tatar, and J. Van Zile. 1993. *Hula Pahu: Hawaiian Drum Dances*. Honolulu: Bishop Museum Press.

Kahoʻolawe Island Reserve Commission and State of Hawaiʻi. 2014. *Kahoʻolawe Island Reserve FY14 Year in Review.*

Kamakau, S. M. 1964. "Ka poʻe kahiko = The People of Old." In *Bernice P Bishop Museum Special Publication* 51, edited by D. B. Barrère, ix. Honolulu: Bishop Museum Press.

———. 1976. "The Works of the People of Old = Na hana a ka poʻe kahiko." In *Bernice P Bishop Museum Special Publication* 61, edited by D. B. Barrère, viii. Honolulu: Bishop Museum Press.

———. 1991. "Tales and Traditions of the People of Old = Na moolelo o ka poe kahiko." In *Bernice P Bishop Museum Special Publication* 94, edited by D. B. Barrère. Honolulu: Bishop Museum Press.

———. 1992. *Ruling Chiefs of Hawaii.* Rev. ed. Honolulu: Kamehameha Schools Press.

Kameʻeleihiwa, L. 1992. *Native Land and Foreign Desires: How Shall We Live in Harmony? = Ko Hawaiʻi ʻāina a me nā koi puʻumake a ka poʻe haole: pehea lā e pono ai?* Honolulu: Bishop Museum Press.

Kanahele, G. H. S. 1986. *Kü Kanaka Stand Tall: A Search for Hawaiian Values.* Honolulu: University of Hawaiʻi Press and Waiaha Foundation.

Kanahele, K. H. 2014. "Kanaloa." In *Hawaiian Studies 102: Hawaiʻi Spirituality.* Hawaiʻi Community College.

———. 2015. "Haumea." In *Hawaiian Studies 102: Hawaiʻi Spirituality.* Hawaiʻi Community College.

Kanahele, P. 2005. "I Am This Land, and This Land Is Me." *Hulili* 2 (1): 21–30.

———. 2014. "Introduction to Kū." In Papakū Makawalu Workshop. Hilo: University of Hawaiʻi.

Kanahele, P. K. 2001. *Holo Mai Pele.* Honolulu: Pacific Islanders in Communications, Edith Kanakaʻole Foundation. Distributed by Native Books.

———. 2011. *Ka Honua Ola: ʻEliʻeli Kau Mai (The Living Earth: Descend, Deepen the Revelation).* Honolulu: Kamehameha Publishing.

Kanahele, P. K., H. Kanahele-Mossman, A. K. Nuʻhiwa and K. Kealiʻikanakaʻole. 2009. *Kūkulu Ke Ea a Kanaloa: The Culture Plan for Kanaloa Kahoʻolawe.* Honolulu: Kahoʻolawe Island Reserve Commission.

Kawaharada, D. 2010. "Voyaging Chiefs of Kāneʻohe Bay." Retrieved from http ://www.webcitation.org/5rTg70xI4.

Kealiʻikanakaʻoleohaililani, K. 2013. Several personal conversations with the author on the subject of Hawaiʻi worldview.

———. 2013a. "Creating Meaning from Names." Hawaiʻicology: Embodying the Hawaiʻi Universe series. *Hawaii Landscape*, March/April, 30–31.

———.2013b. "Creating the A-Kua Potential in Landscaping." Hawaiʻicology Embodying the Hawaiʻi Universe series. *Hawaii Landscape*, November/December, 30–31.

———. 2013c. "In the Company of Plant People." Hawai'icology Embodying the Hawai'i Universe series. *Hawaii Landscape*, January/February, 30–31.

———. 2013d. "Kalo Kosmology." Hawai'icology Embodying the Hawai'i Universe series. *Hawaii Landscape*, July/August, 30–31.

———. 2015. "Pele." Video. On *Kumukahi: Living Hawaiian Culture* website, edited by Keola. Kamehameha Publishing. http://www.kumukahi.org/

———. 2016. Hula. Several personal conversations with the author.

Keauokalani, K. 1932. "Kepelino's Traditions of Hawaii." In *Bulletin: Bernice P Bishop Museum* 95, edited by M. W. Beckwith. Honolulu: Bernice P Bishop Museum.

Kent, H. W. 1986. *Treasury of Hawaiian Words in One Hundred and One Categories*. Honolulu: Masonic Public Library of Hawaii. Distributed by University of Hawai'i Press.

Kievit, J. A. 2003. "A Discussion of Scholarly Responsibilities to Indigenous Communities." *American Indian Quarterly* 27: 1, 2.

Kikiloi, K. 2010. "Rebirth of an Archipelago: Sustaining a Hawaiian Cultural Identity for People and Homeland." *Hulili: Multidisciplinary Research on Hawaiian Well-Being* 6: 73–115.

Kitchin, R., and M. Dodge. 2007. "Rethinking Maps." *Progress in Human Geography* 31 (3): 331–344.

Kitchin, R., J. Gleeson, and M. Dodge. 2013. "Unfolding Mapping Practices: A New Epistemology for Cartography." *Transactions of the Institute of British Geographers* 38: 480–496.

Kovach, M. 2009. *Indigenous Methodologies: Characteristics, Conversations, and Contexts*. Toronto: University of Toronto Press.

Lake, J. K. n.d. *Nā hana a me nā 'ike ha'i'ōlelo o nā ali'i a me nā papa kanaka*. Honolulu: Self-published.

Lorde, A. 1984. *Sister Outsider: Essays and Speeches*. Trumansburg, NY: Crossing Press.

Low, S. 2013. *Hawaiki Rising:* Hōkūle'a, *Nainoa Thompson, and the Hawaiian Renaissance*. Waipahu, HI: Island Heritage Publishing.

Luomala, K. 1965. "Creative Processes in Hawaiian Use of Place Names in Chants." In *Lectures and Reports 4th International Congress for Folk-Narrative Research in Athens, 1964. Laographia* 22: 234–237.

Makemson, M. W. 1939. "Hawaiian Astronomical Concepts II." *American Anthropologist* 41 (4): 589–596.

Malo, D. 1971. *Hawaiian Antiquities = (Moolelo Hawaii)*. 2nd ed. Honolulu: Bishop Museum Press.

Mark, D. 1993. "Human Spatial Cognition." In *Human Factors in Geographical Information Systems*, edited by D. Medyckyj-Scott and H. M. Hearnshaw, 51–60. London: Belhaven Press.

McDougall, B. N. 2014. "Putting Feathers on Our Words: Kaona as a Decolonial Aesthetic Practice in Hawaiian Literature." *Decolonization: Indigeneity, Education & Society* 3 (1): 1–22.

McMaster, R., and S. McMaster. 2002. "A History of Twentieth-Century American Academic Cartography." *Cartography and Geographic Information Science* 29 (3): 305–321.

Meyer, M. A. 2003a. Email with author.

———. 2003b. *Ho'oulu—Our Time of Becoming: Hawaiian Epistemology and Early Writings*. Honolulu: 'Ai Pohaku Press.

Mihesuah, D. A., ed. 1998. *Natives and Academics: Researching and Writing about American Indians*. Lincoln: University of Nebraska Press.

Mitchell, T. 1988. *Colonising Egypt*. Berkeley: University of California Press.

Montello, D. R. 1997. "Unit 006: Human Cognition of the Spatial World." Retrieved from http://www.ncgia.ucsb.edu/education/curricula/giscc/units/u006/u006.html.

Mundy, B. 1998. "Mesoamerican Cartography." In *Cartography in the Traditional African, American, Arctic, Australian, and Pacific Societies*, edited by D. Woodward and G. M. Lewis, 183–256. Chicago: University of Chicago Press.

Murphy, K. 2015. "On the Water." In *Never Lost: Navigation*, edited by R. J. Semper. http://www.exploratorium.edu/neverlost/#/navigation/on_the_water.

Nietschmann, B. 1995. "Defending the Miskito Reefs with Maps and GIS: Mapping with Sail, Scuba, and Satellite." *Cultural Survival Quarterly* 18 (4). https://www.culturalsurvival.org/publications/cultural-survival-quarterly/colombia/defending-miskito-reefs-maps-and-gps-mapping-sail-.

Nogelmeier, P. 2001. "Introduction." In *He Lei no 'Emalani: Chants for Queen Emma Kaleleonālani*, 1–7. Honolulu: Queen Emma Foundation, Bishop Museum Press.

———. 2010. "The Hawaiian Language." In *Long Story Short*, TV series, edited by L. Wilcox. Honolulu: PBS Hawai'i.

Norton-Smith, T. M. 2010. *The Dance of Person and Place: One Interpretation of American Indian Philosophy*. Albany: State University of New York Press.

Nu'uhiwa, K. 2011. "Haumea: Establishing the Sacred Space." Presentation at Papakū Makawalu workshop. Keauhou, Kona, Hawai'i.

Office of Hawaiian Affairs. 2013. Kīpuka Database. Retrieved from http://kipukadatabase.com/.

'Ōiwi TV. 2014. "World Wide Voyage, Crew Profile: Ka'iulani Murphy." In *Hōkūle'a Worldwide Voyage*, 6:50. YouTube.

Oliveira, K.-A. R. K. N. 2014. *Ancestral Places: Understanding Kanaka Geographies*. Corvallis: Oregon State University Press.

Olson, J. 1981. "Spectrally Encoded Two Variable Maps." *Annals of the Association of American Geographers* 71 (2): 259–276.

Perreira, H. K. 2011. "He Haʻi ʻŌlelo Kuʻuna: Nā Hiʻohiʻona me nā Kiʻina Hoʻāla Hou i ke Kākāʻōlelo." PhD diss., Hawaiian and Indigenous Language and Culture Revitalization, University of Hawaiʻi, Hilo.

Poepoe, J. M. 1906. "Moʻolelo Hawaiʻi Kahiko." *Ka Naʻi Aupuni*, September 25–28, 1906.

Polynesian Voyaging Society. 2012a. "Estimating Distance and Direction Traveled." Retrieved from http://pvs.kcc.hawaii.edu/ike/hookele/estimating_distance_direction.html.

———. 2012b. "Estimating Position." Retrieved from http://pvs.kcc.hawaii.edu/ike/hookele/estimating_position.html.

———. 2012c. "Holding a Course." Retrieved from http://pvs.kcc.hawaii.edu/ike/hookele/holding_a_course.html.

———. 2012d. "Locating Land." Retrieved from http://pvs.kcc.hawaii.edu/ike/hookele/locating_land.html.

———. 2014a. "The Celestial Sphere." Retrieved from http://www.hokulea.com/education-at-sea/polynesian-navigation/polynesian-non-instrument-wayfinding/the-celestial-sphere/.

———. 2014b. "Hawaiian Star Lines." Retrieved from http://www.hokulea.com/?s=star+lines.

———. 2014c. "Polynesian Wayfinding." Retrieved from http://www.hokulea.com/education-at-sea/polynesian-navigation/polynesian-non-instrument-wayfinding/.

Pukui, M. K. 1949. "Songs (Meles) of Old Kaʻu, Hawaii." *Journal of American Folklore* 62 (245): 247–258.

———. 1983. *ʻŌlelo noʻeau: Hawaiian Proverbs and Poetical Sayings*. Honolulu: Bishop Museum Press.

———. n. d. "How Legends Were Taught." Unpublished paper.

Pukui, M. K., and S. H. Elbert. 2003. *Hawaiian Dictionary: Hawaiian-English, English-Hawaiian*. Honolulu: University of Hawaiʻi Press. http://www.ulukau.org/elib/cgi-bin/library?c=ped.

Pukui, M. K., S. H. Elbert, and E. T. Moʻokini. 1974. *Place Names of Hawaiʻi: Revised and Expanded Edition*. Honolulu: University of Hawaiʻi Press.

Pukui, M. K., E. W. Haertig, and C. A. Lee. 1972. *Nānā i ke kumu = Look to the Source*. Honolulu: Hui Hanai.

Pukui, M. K., and A. L. Korn. 1979 (1973). *The Echo of Our Song*. Honolulu: University of Hawaiʻi Press.

Reiny, S. 2011. "Performing Arts." *Hana Hou* 14 (2) (April–May).

Roberts, L. 2012. "Mapping Cultures: A Spatial Anthropology." In *Mapping Cultures: Place, Practice, Performance*, edited by L. Roberts, 1–25. Basingstoke, UK: Palgrave Macmillan.

Robinson, A. H., and B. B. Petchenik. 1976. *The Nature of Maps: Essays toward Understanding Maps and Mapping*. Chicago: University of Chicago Press.

Rodaway, P. 1994. *Sensuous Geographies: Body, Sense, and Place*. London: Routledge.

Rundstrom, R. A., and D. Deur. 1999. "Reciprocal Appropriation: Toward an Ethics of Cross-Cultural Research." In *Geography and Ethics*, edited by J. D. Proctor and D. M. Smith, 237–250. London: Routledge.

Sahota, P. C. 2009. "Research Regulation in American Indian/Alaska Native Communities: Policy and Practice Considerations." NCAI Policy Research Center, http://depts.washington.edu/ccph/pdf_files/Research%20Regulation%20in%20AI%20AN%20Communities%20-%20Policy%20and%20Practice.pdf.

Silva, N. K. 2004. *Aloha Betrayed: Native Hawaiian Resistance to American Colonialism*. Durham, NC: Duke University Press.

———. 2017. *The Power of the Steel-Tipped Pen: Reconstructing Native Hawaiian Intellectual History*. Durham, NC: Duke University Press.

Smelser, N. J., and P. B. Bates. 2001. *International Encyclopedia of the Social and Behavioral Sciences*. Oxford, UK: Pergamon Press.

Smith, L. T. 1999. *Decolonizing Methodologies: Research and Indigenous Peoples*. New York: Zed Books. Distributed in the US by St. Martin's Press.

Snow, J. 2014. "Beyond the Binary: Portraits of Gender and Sexual Identities in the Hawaiian Community." *Mana* 2 (5): 22–28.

Soja, E. W. 1989. *Postmodern Geographies: The Reassertion of Space in Critical Social Theory*. London: Verso.

Stagner, I. W. 2011. *Kumu Hula Roots and Branches*. Honolulu: Island Heritage Publishing.

Steinhauer, E. 2002. "Thoughts on an Indigenous Research Methodology." *Canadian Journal of Native Education* 26 (2): 69–81.

Stillman, A. K. 2003. "On Mele: The Heart of Hula." In *American Aloha: Hula Beyond Hawai'i*, edited by PBS. Online overview of broadcast film, premier August 5.

———. 2005. "Textualizing Hawaiian Music." *American Music* 23 (1): 69–94.

Sutton, P. 1998. "Icons of Country: Topographic Representations in Classical Aboriginal Traditions." In *Cartography in the Traditional African, American, Arctic, Australian, and Pacific Societies*, edited by D. Woodward and G. M. Lewis, 353–386. Chicago: University of Chicago Press.

Tabag, J. N. 2010. *Māhūwahine, he ola kupaianaha!: (Māhūwahine, Hawaiian Transsexual Experiences)*. Honolulu: University of Hawai'i at Manoa.

Tanahy, D. 2006. "Kapa Making and Processing." Retrieved from http://www.kapahawaii.com/how-to-make-hawaiian-tapa.html.

Tartar, E. 1982. "Nineteenth Century Hawaiian Chant." In *Pacific Anthropological Records No. 33*, edited by Department of Anthropology, Bernice Pauahi Bishop Museum. Honolulu: Bishop Museum Press.

Teves, S. N. 2015. "Tradition and Performance." In *Native Studies Keywords*, edited by S. N. Teves, A. Smith, and R. Michelle. Tucson: University of Arizona Press.

Thompson, N. 2012. "Recollections of the 1980 Voyage to Tahiti." Retrieved from http://pvs.kcc.hawaii.edu/holokai/1980/nainoa_to_tahiti.html.

Tobias, T. N. 2000. *Chief Kerry's Moose: A Guidebook to Land Use and Occupancy Mapping, Research Design, and Data Collection*. Vancouver: Ecotrust Canada, Union of BC Indian Chiefs.

———. 2009. *Living Proof: The Essential Data-Collection Guide for Indigenous Use-And-Occupancy Map Surveys*. Vancouver: Ecotrust Canada, Union of BC Indian Chiefs.

Tsuha, K. 2008. "Kaulana Mahina, The Hawaiian Lunar Calendar." In *Puana Ka 'Ike Lecture Series*. Kamuela, HI: The Kohala Center.

Tuan, Y.-f. 1990. *Topophilia: A Study of Environmental Perception, Attitudes, and Values*. New York: Columbia University Press, Morningside.

———. 2001. *Space and Place: The Perspective of Experience*. Minneapolis: University of Minnesota Press.

Turnbull, D. 2003. *Masons, Tricksters, and Cartographers: Comparative Studies in the Sociology of Scientific and Indigenous Knowledge*. London: Routledge.

Veary, N. 1989. *Change We Must: My Spiritual Journey*. Honolulu: Institute of Zen Studies.

Wilcox, C., V. Hollinger, K. Hussey, and P. Nogelmeier. 2003. *He Mele Aloha: A Hawaiian Songbook*. Honolulu: 'Oli'Oli Productions.

Wilmshurst, J. M., T. L. Hunt, C. P. Lipoc, and A. J. Anderson. 2011. "High-Precision Radiocarbon Dating Shows Recent and Rapid Initial Human Colonization of East Polynesia." *Proceedings of the National Academy of Sciences (PNAS)* 108 (5): 1815–1820.

Winichakul, T. 1994. *Siam Mapped: A History of the Geo-Body of a Nation*. Honolulu: University of Hawai'i Press.

Wong-Kalu, H. 2001. "Hina." In *'O au no keia: Voices from Hawai'i's Mahu and Transgender Communities*, edited by A. Matzner, 217–228. United States: Xlibris.

Woodward, D., and G. M. Lewis. 1998. *Cartography in the Traditional African, American, Arctic, Australian, and Pacific Societies*. Vol. 2, book 3, in The History of Cartography series, edited by University of Chicago Press, xxi. Chicago: University of Chicago Press.

Index